1992

Mathematical modelling

A case study approach

Mathematical modelling

A case study approach

Dick Clements
University of Bristol

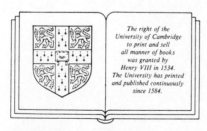

The right of the
University of Cambridge
to print and sell
all manner of books
was granted by
Henry VIII in 1534.
The University has printed
and published continuously
since 1584.

Cambridge University Press
Cambridge
New York   Port Chester
Melbourne   Sydney

Published by the Press Syndicate of the University of Cambridge
The Pitt Building, Trumpington Street, Cambridge CB2 1RP
40 West 20th Street, New York NY 10011, USA
10 Stamford Road, Oakleigh, Melbourne 3166, Australia

First published 1989

Printed in Great Britain at the University Press, Cambridge

British Library cataloguing in publication data

Clements, Dick
Mathematical modelling.
1. Mathematical models. Applications
I. Title
511′.8

Library of Congress cataloguing in publication data

Clements, Dick.
Mathematical modelling.
Bibliography: p.
1. Mathematical models. I. Title.
QA401.C55   1989   00.4′34   88-37587

ISBN 0 521 34340 2

Contents

Preface   *page* vii

PART I

1 Introduction   1

2 The background to mathematical modelling   6
2.1   The need to teach mathematical modelling   6
2.2   Early expositions of the methodology of modelling   10
2.3   The development of iterating methodologies   12
2.4   Methodologies of modelling and the scientific method   17
2.5   The systems movement and Checkland's methodology   18
2.6   A complex linkage methodology of mathematical   20
        modelling
2.7   What use is a methodology anyway?   23
2.8   Conclusion   23

3 Mathematical modelling in practice   24
3.1   Understanding the problem   24
3.2   Iterative refinement of models   28
3.3   Mathematical constraints on models   33
3.4   The importance of choice of notation   37
3.5   Data and sensitivity of conclusions   39
3.6   Use of mathematical tools   40
3.7   Conclusion   40

4 A course of case studies   41
4.1   The context of the case study course   41
4.2   The need to teach mathematical modelling   43
4.3   The objectives of teaching modelling   46
4.4   Simulations, case studies and the hybrid concept   46
4.5   The creation of the case studies   50

4.6   Practical considerations for the course   52
4.7   Student reactions to the course   54
4.8   The demands made on tutors   55
4.9   Assessment of student performance   56
4.10   Other similar developments   58

PART II

5   **Andover Aerospace Components**   59
5.1   The source documents   60
5.2   General lines of donor's solution   71
5.3   Commentary on student solutions   74

6   **The Leather Working Machinery Group**   76
6.1   The source documents   77
6.2   General lines of donor's solution   79
6.3   Commentary on student solutions   82

7   **British Knitted Garments Group**   85
7.1   The source documents   86
7.2   General lines of donor's solution   97
7.3   Commentary on student solutions   101

8   **Gamma Avionics**   104
8.1   The source documents   105
8.2   General lines of donor's solution   110
8.3   Commentary on student solutions   113

9   **Fanning Industries**   117
9.1   The source documents   118
9.2   General lines of donor's solution   123
9.3   Commentary on student solutions   129

10   **Winchester Aircraft Company**   132
10.1   The source documents   133
10.2   General lines of donor's solution   138
10.3   Commentary on student solutions   141

11   **Natural Gas**   143
11.1   The source documents   144
11.2   General lines of donor's solution   146
11.3   Commentary on student solutions   156

References   159
Subject index   163
Name Index   165

Preface

In 1973, when I joined the staff of the University of Bristol as a lecturer in the Department of Engineering Mathematics, I found my new colleagues in the preliminary stages of planning a novel degree course to be known as Engineering Mathematics. One of the primary objectives of this degree course was to produce graduates with not only a sound mathematical education but also the ability to apply their mathematical knowledge to the solution of the problems of commerce and manufacturing industry. As a contribution to the design of this course I proposed to my colleagues that we should include, in the overall course structure, a course of practical mathematical activities aimed at introducing students to a range of industrially relevant mathematical and para-mathematical activities. Thus was born the Case Study course.

The concept of the course called for the collection of resource material from a wide cross-section of industry and commerce. To support this phase of the work a grant was sought from the Nuffield Foundation under what was then known as their Small Grants Scheme for Undergraduate Teaching. The grant was approved and the task of gaining the co-operation of industry and commerce in the creation of the resource material for the course proceeded throughout 1976 and 1977. The course itself was first used with undergraduate students in early 1978 and has been run annually since then. It has become one of the distinctive features of the Engineering Mathematics course and is, incidentally, highly popular with the students.

This book is divided into two sections. Part I (Chapters 1 to 4) introduces the topic of mathematical modelling and places it in its context *vis-à-vis* the Case Study course. It is my opinion that an ability to create and use mathematical models is an essential skill for an effective user of mathematics in the industrial or commercial context. That alone does not, however, suffice. There are a range of other mathematical and para-mathematical abilities which are also needed. Not least amongst these is the ability to communicate mathematical ideas both to other mathematicians and to those with less mathematical skill or inclination. The Case Study method aims not only to teach a modicum of mathematical modelling but also to develop these other skills.

Part II (Chapters 5 onward) introduces a number of the individual case studies which have been used in the Case Study course. Each of these chapters is divided into three sections consisting of, respectively, the source material which is used to present the case study to the students, an outline of the work done by the staff of the company or institution in which the problem originally arose, and some comments about the methods which nine generations of students have used to 'solve' the problems.

There are many individuals and corporate bodies without whose goodwill and assistance this course would not have existed and this book would not have been written. My thanks must, naturally, go firstly to the Nuffield Foundation, without whose support the creation of the course would not have been possible. Thanks are also due to my colleagues in the Engineering Mathematics Department, particularly to Prof Ronald Milne for his continual and continuing encouragement, help and advice, and to Dr Bruce Pilsworth, who has, since 1983, shared in the running of the course. To these I must add all my friends and colleagues in Universities, Polytechnics, and equivalent institutions in many parts of the world who have shown, through their enthusiastic discussion of this work and in myriad other ways, their interest and encouragement. My gratitude is also due to the 'victims', the students of Engineering Mathematics at Bristol over the years, for their (in most cases) ready acceptance of an unusual and novel learning experience and their eventual enthusiasm for the medium and delight in their newly acquired skills, and to Louise Clements, who assisted in the administration of the resource collection phase of the project during 1976 and 1977.

I must also thank the Military Aircraft Division of British Aerospace plc at Warton, British Gas plc, Courtaulds plc, The Marconi Research Centre of GEC Research Limited, Mirrlees Blackstone (Stockport) Limited and a number of other companies who have generously contributed to the course and given their permission for the material provided to be included in this book. In particular I am most grateful to those employees of the afore-mentioned companies through whose agency I received the companies' assistance. I thank them for their cheerful helpfulness and the tolerance with which they endured my persistent enquiries. Finally my thanks go to my colleague and friend Margaret Irish for the interest, encouragement and tolerance she has shown during the writing of this book.

April 1988   DICK CLEMENTS

PART I

1

Introduction

This book is the culmination of more than ten years work by the author on ways of teaching undergraduate students how to become not only competent mathematicians but also skilled users of mathematics in the solution of problems arising in the real world.

In a degree course, at least in mathematics and those related disciplines in which mathematics is a tool, three things must be taught. Firstly, there is a body of factual knowledge and technique which students must acquire. Secondly, the skill of extending their mastery of factual knowledge and technique must also be acquired, that is they must learn how to learn more mathematics as and when the need arises. Thirdly, since all the mathematical knowledge and technique in the world is of little use to the practical mathematician, engineer or scientist if the skill of applying that knowledge to their professional problems is missing, students must learn how to use their mathematical knowledge in solving the problems of the real world. In the past the second and third of these aspects of mathematics have not been formally taught, rather it was assumed that students would acquire them incidentally (one might almost say accidentally) whilst studying the facts, theorems and techniques of mathematics. Indeed there has been considerable doubt as to whether these skills could be taught effectively at all. At the same time, though, one of the classical justifications advanced for teaching mechanics has always been that, through the study of mechanics, students will learn something about the art of applying mathematical theory to the problems of the real world. However, it is now increasingly accepted that, whilst it may be extremely difficult if not impossible to learn these skills merely by attending lectures about them (as indeed it is difficult to learn to ride a bicycle by attending lectures on bicycle riding), they can be more or less successfully learnt by practical activity (as can riding a bicycle).

The author's explorations of means of teaching both the second and third aspects of mathematics mentioned above have resulted in two innovations in his own teaching. Firstly, the use of the 'guided reading' method as a means of helping students to develop the skills of extending their mathematical knowledge has been explored. Work relating to this innovation is reported in UMTC (1978) and Clements and Wright (1983). Secondly, to meet the requirement to teach the skills of applying mathematics to the problems of the real world a series of exercises, termed 'case studies', has been created. It is the aim of this book to explain both the underlying principles of these case studies and how they have been used to help students learn the art of applying their mathematical knowledge to the problems of the real world.

The object of these case study exercises, then, is to give undergraduates some experience of applying their mathematical knowledge to the sort of problems that arise in industry and commerce. Each of these case studies is based on a problem which has, at some time, faced an actual company or organisation. The material on which the studies are based was contributed by those organisations together with an outline of the way in which those originally faced with the problem attempted to solve it. Each chapter in part II of this book presents, in a realistic way, a problem situation facing a mythical organisation in the solution of which mathematical skills may assist. The problems are presented in the form of reconstructed design drawings, correspondence, memoranda and reports. Problems in the real world do not, of course, arise neatly packaged and expressed in mathematical notation. Instead they often arise in messy, confused ways and usually expressed in someone else's terminology (for instance chemical or engineering terminology). This confusion is reflected in the presentation of the case studies via the reports, memoranda and other documents.

Each of the chapters in part II of the book also contains an outline of the approach to solving the problem which was adopted by the original organisation which faced the problem and a commentary on the variations on these approaches which have been produced by student groups studying these problems over the years. It is important to realise that the approach adopted by the donor organisation is not necessarily 'correct' or 'ideal'. In the real world solutions are rarely 'correct' and not often 'ideal'. Most real problems admit of a variety of solutions. The solution method chosen for any given problem must be one that is appropriate under the circumstances prevailing. The relevant circumstances may include non-mathematical constraints as well as mathematical considerations. Problem solving is a much more open-ended activity than theorem proving.

In part I of the book some more general material on mathematical modelling and the use of the Case Study course at Bristol University is presented. It is hoped that this material will assist both lecturers wishing to use the material in part II in their own teaching and students wanting to learn more about applying mathematics to the problems of the real world.

It is appropriate, at this point, to say something more detailed about the ways in which this book might be used. This depends, of course, on who you, the reader, are and on what you hope to learn from the book. Firstly, let us consider what use an individual student of mathematics (and let us not restrict this category to undergraduates but include students of all ages and levels of experience) might make of the book. If you are such a student it would be possible for you merely to read this book much as you would any other book. If you restrict yourself to this however, you will miss much of the benefit that is to be derived from the book, and incidentally much of the pleasure and satisfaction! Ideally you should use this book in association with a structured course in which you will work with others on the solution of the problems in part II of the book. In this way you will enjoy the same benefits that those working in industry and commerce enjoy – the opportunity to discuss with your fellows the advantages and disadvantages of the various possible approaches to the problem, the joint decision taking about which avenue to pursue and the mutual support and stimulus which is derived from common endeavour. It goes without saying, of course, that the benefit you derive from such problem solving will be greatly reduced if you read the outline of the donor's solution before attempting your own. Such prior knowledge will almost certainly constrain your own thinking and implicitly limit your potential creativity. By attempting your own solution of the problem you will be stimulated not only to work towards an appropriate solution but also to consider and evaluate a range of possible ways of obtaining that solution.

If, through force of circumstance, you have to use this book as an isolated individual you will still derive more benefit from thinking out for yourself a solution which answers the problem than you will by merely reading a solution provided by someone else. In many ways it is true to say of this book that the correctness of the answers you obtain is less important than the experience of selecting or devising mathematical ways of answering the problems posed.

Whether you are tackling each problem on your own or as a member of a group, taking a course in an educational institution or reading this book at home for general interest, you will obtain much benefit from the struggle to apply your mathematical knowledge to these novel situations. You should anticipate that this will be, at first, a disorientating and perhaps, to a certain extent, a frightening situation. Do not be disheartened but persevere and you will soon be amply rewarded by the satisfaction of finding yourself able to enhance your understanding of the world in which you live and work through the application of mathematical principles and techniques.

So far we have considered the uses students might make of this book. If, on the other hand, you are a teacher or lecturer seeking stimulation, guidance or just material for a course in the application of mathematics you will find that this book addresses your needs also. Obviously the case studies described in part II of the book can be incorporated into your courses and the outline of the original solution adopted and the commentary on likely or possible student variations can be used in your courses in whatever ways you see fit. In chapter

4 there is an extensive description of the course for which this material was originally developed and within which it has been used successfully for a number of years. Chapters 2 and 3 contain less specific material relating to mathematical modelling in general. In order to understand why this material is included let us return to the analogy between learning to be a competent practitioner of mathematical modelling and learning to ride a bicycle. It was suggested above that these two share the property that neither is easily learnt by attending lectures on the theory of the activity and that both are best learnt by practice. The analogy between bicycle riding and mathematical modelling may be extended to include the role of the teacher. Whilst the learner bicycle rider may only acquire the desired expertise through practice, teachers of bicycle riding will be much more effective if they understand something of the theory of bicycle riding. Such knowledge enables them to analyse the performance of the learner and give appropriate advice and instruction, reinforcing and encouraging appropriate behaviours in the learner and correcting inappropriate ones. This is not to say that those without the theoretical understanding of bicycle riding are not capable of teaching it, but their effectiveness in so doing is reduced by their lack of knowledge. So too with mathematical modelling – experienced practitioners of the art may be capable of passing on their art and skill in an instinctive way but those who wish to be the most effective teachers of the subject must not only have experience of the art and practice of mathematical modelling but must also possess a deeper understanding of the subject and have appropriate mental models of the process which they desire to teach.

Again, whilst practice is the best (and probably the only) way to take the initial steps in learning bicycle riding, once the rider has acquired a minimal practical skill advancement to the highest levels of practical skill in riding is aided by theoretical knowledge of the mechanics and dynamics of bicycles. Perhaps this assertion seems a little fanciful in respect of a relatively simple activity like bicycle riding – if so then translate bicycle riding to driving a motor vehicle. The learner driver acquires his or her initial skill by practical means but, before the driving test can be tackled, some theoretical knowledge is needed to illuminate the best way of using, on the roads, the practical skills which have been acquired. Drivers who wish to advance to passing the test of the Institute of Advanced Motorists, or to reach the even higher standards of the Police traffic patrol driver or the qualified chauffeur, will certainly need further theoretical knowledge of the subject as well as additional practical skill. The analogy carries through to mathematical modelling. After acquiring a basic skill in modelling through practical activities, those who wish to become more expert are helped on the route by some study of the theory of modelling as well as more advanced practice and experience. Such theoretical knowledge helps them to place their practical activities in context and to guard against suboptimal or dysfunctional behaviour.

From this analogy it may be concluded that, at some time or other, both teachers and students of mathematical modelling need to study the theoretical background to, and methodologies of, modelling as well as the practicalities of the subject. To meet these needs chapter two presents an analysis of the background to mathematical modelling, including some comments on the history of the subject and an outline of the development of methodologies of modelling, and chapter three attempts, insofar as such an attempt is possible, to illuminate the practicalities of modelling. This attempt takes the form of something akin to thinking aloud, that is several simple models are developed and comments are made on how and why various decisions were taken during the development process.

In these ways this book attempts to satisfy the needs of both teachers and students of mathematics for assistance in developing those skills which experienced users of mathematics routinely deploy whilst using their mathematical knowledge to illuminate their understanding of the world.

2

The background to mathematical modelling

In chapter one it was suggested that a degree course in mathematics should seek to teach firstly a body of mathematical knowledge, secondly the ability to extend that knowledge independently and thirdly the ability to use that knowledge. It was also suggested that mathematics courses in tertiary education have traditionally concentrated on the first of these to the detriment of the other two. There is however, within the discipline of mathematical education today, an identifiable group of teachers and lecturers who, to a greater or lesser extent, believe that the teaching of mathematical modelling is vital to the development of an ability to use mathematics and that such teaching is necessarily a distinct activity from the teaching of other topics in mathematics. The existence of such a group, who might be characterised as the mathematical modelling movement, is a distinctively recent innovation in mathematics – it would have been difficult, for instance, to identify any such grouping prior to the mid nineteen sixties. We might ask what circumstances have led to the birth of the movement and why has its development been so rapid? In this chapter the origins of the movement will be described, the development of the theory of mathematical modelling traced and some new insights into the nature of the mathematical modelling process proposed.

2.1 The need to teach mathematical modelling

A panel discussion, organised in 1961 by the [American] Society for Industrial and Applied Mathematics, between Professors Carrier, Courant, Rosenbloom and Yang entitled 'Applied mathematics: what is needed in research and education' is reported by Greenberg (1962). Greenberg, who chaired the meeting, referred, in his introduction, to a revolution in mathematics teaching in the USA and to the need, in the light of this, to determine a design for the future of applied mathematics. It is particularly interesting to notice what skills the distinguished panellists identified as essential for applied mathematicians. In his address Prof Carrier said

To [contribute to the quantitative understanding of scientific phenomena], [applied mathematicians] must be so thoroughly informed in the fundamentals of some broad

6

segment of the sciences . . . that they can pose the question or family of questions they pursue as a mathematical query using, as the occasion demands, either time-honoured and well established scientific laws . . . or carefully conjectured models. Such an applied mathematician must also have an understanding of mathematics, a knowledge of technique, and such skill that he can use either rigorously founded techniques or heuristically motivated methods to resolve the mathematical problem, and he must do so with a full realisation as to the implications of each with regard to reliability and interpretation of results. In particular, the applied mathematician must be very skillful at finding that question (or family of questions) such that the answer will fill the scientific need while the extraction of the answer and its interpretation are not prohibitively expensive.

Prof Carrier went on to emphasise that, in his opinion, such skills could only be inculcated in mathematics students if their study of mathematics was closely integrated with study of the sciences, engineering and other disciplines in which that mathematics found its applications.

Prof Rosenbloom voiced the opinion that even one of the weightiest and most popular of classical mathematical physics textbooks was both too much and too little for a complete professional training for a physicist. He described it as too little in the sense that

. . . there is nothing in there about how you would actually set up a problem.

He subsequently added

The basic problem is that applied mathematics is an art in which mathematics is only a part. We have a situation in the real world from which you have to create a mathematical model by idealisation and simplification. . . . We then study the mathematical model using all the power and technique of mathematics on that, often using our intuition from the interpretation that we had in mind. Then the test of our model is whether, when you interpret it back in reality, it works. And the middle part, the study of this mathematical model, which is the game of the mathematician, is only part of the whole process of applied mathematics.

In all four contributions to this discussion there is to be found an appreciation of the importance, for applied mathematics and its fields of application, both traditional and modern, of modelling and model building, and a realisation that, at the time of the discussion, something needed to be done to improve the teaching of these skills.

Ten years later Prager, in the introductory remarks to a 'Symposium on the future of applied mathematics' (Prager, 1972), described the applied mathematician thus

With the pure mathematician, the applied mathematician shares the interest in developing new mathematics, and with the scientist or engineer, the interest in applying mathematics to the improvement of our understanding and control of natural or man-made environments. As an intermediary between these groups, the applied mathematician should appreciate, though not necessarily emulate, the pure mathematician's insistence on rigour as well as the willingness of scientists and engineers to accept heuristic reasoning. He must be able to construct, not only a rigorous proof of a mathematical proposition, but also a workable mathematical model of the phenomenon he plans to investigate.

Very often, the applied mathematician's skill in the construction of suitable models will contribute as much to the success of an investigation as his knowledge of the analytical or numerical techniques that will be needed to treat the mathematical relations governing the behaviour of the adopted model. His familiarity with these techniques enables him to foresee difficulties that may result from the inclusion of certain effects in the model. He will then question his clients about the need for including these effects, warn of the mathematical consequences of this inclusion, and stress the fact that a coarse model that is readily manipulated mathematically may yield a better insight into a natural phenomenon or technical process than a more refined but mathematically unwieldy model.

Having given this description of the work of applied mathematicians, he went on to suggest how their education should be changed in order to better fit students for the coming demands.

I believe that, in teaching applied mathematics, we should devote more time to the process of discovery than is customary today. This will not be easy on account of the tradition that requires us to present our results in an orderly and logical way and makes us reluctant to report the often erratic and illogical ways in which these results were obtained. We should overcome this reluctance because an account of the process of discovery will frequently be more useful to the student of applied mathematics than the particular result.

We see that Prager, like his colleagues, appreciated the demands which the applied mathematician faced (and indeed still faces) and recognised that some changes were needed in the education of applied mathematicians to fit them better to deal with these. His prescription for these changes goes some way to meeting the demand, but the proponents of mathematical modelling would argue that it is even better for students to experience the process of mathematical discovery than to hear an account of it.

In 1970 McLone undertook a major study of the perceptions of both the employers of mathematics graduates and the graduates themselves of the relevance and usefulness of mathematics degree courses in UK universities. The study was conducted through the medium of postal surveys both of employers and of graduates from mathematics departments of UK universities in the years 1964, 1966 and 1968. The findings of the survey are reported in Griffiths and McLone (1971) and in McLone (1973). Amongst the conclusions drawn from the responses were

A transfer in emphasis to the basic problems of mathematical modelling and the formulation of mathematical problems is both necessary and desirable, if a mathematician is to be of benefit to the environment in which he works after graduation.

and

The request most often made (by employers) is for mathematics graduates with an appreciation of the applicability of their subject in other fields and an ability to express problems, initially stated in non-mathematical terms, in a form amenable to mathematical treatment with the subsequent re-expression in a readily understandable form to non-mathematical colleagues. This request is usually accompanied by a statement which indicates that this is not often found in those currently graduating from British universities.

and

. . . an important aspect of applied mathematics which is emphasised by all groups is mathematical modelling, that is, the modelling of real situations in mathematical terms. (McLone, 1973)

Writing as an industrial employer of mathematics graduates, Klamkin (1971) described what he saw as the role of the mathematician in industry. He stated

After a problem is recognised, the next stage is the formulation of something precise to work on. This consists of making a mathematical model of the physical situation which, on the one hand, has to be simple enough to permit a complete mathematical analysis but, on the other hand, is sufficiently close to reality to be relevant to the actual physical problem being considered. This model building is probably the most difficult and valuable task for the industrial mathematician.

Klamkin went on to criticise the abilities of graduates of both bachelors and doctoral degree courses in these areas.

Gaskell and Klamkin (1974) reported the findings of an informal survey carried out by the Mathematical Association of America in which the views of heads of non-academic groups of mathematicians (such groups being, in fact, primarily industrial) were sought on a number of issues related to the employment of mathematics graduates in their groups. Included in these issues were some questions relating to the adequacy and appropriateness of their education. The paper was comprised mainly of quotations from the responses received, these being collated and arranged according to issues. A number of these quotations made points relevant to a discussion of mathematical modelling. In particular the comment

Training could be improved by teaching courses in the formation of mathematical models – probably by showing students how to describe mathematically and try to optimize some of the non-mathematical activities that they are familiar with in their everyday life . . .

and a closing comment by the authors

Our traditional educational procedures have provided only training in mathematical manipulation, with extremely little attention to problem recognition, formulation, and follow-up.

are relevant to our concerns here.

Taken as a whole the preceding comments seem to indicate that there was, in the sixties and early seventies, a growing recognition that an identifiable body of skills, loosely characterised as an ability in mathematical modelling, existed and was an important acquisition for applied mathematicians whether intending to follow careers in the academic world or in industry. There was, further, an acknowledgement that traditional degree courses did not succeed nearly well enough in inculcating these skills and that some action was necessary, at tertiary and also at lower levels in the educational system, to rectify this situation. The mathematical modelling movement emerged in response to this need.

One indication of the emerging movement was the increasing number of publications, concerned with the philosophy of applied mathematics and related subjects, in which modelling was mentioned as a desirable or essential objective or component of applied mathematics education; for example, Lin (1967, 1976, 1978), Woods (1969), Ford and Hall (1970), Hall (1972) and Lighthill (1979). More recently there has been a considerable body of published literature concerned with mathematical modelling and a number of journals have been introduced covering the field (the *UMAP Journal*, the *Journal of Mathematical Modelling for Teachers* and its successor, the journal *Teaching Mathematics and Its Applications* published by the Institute of Mathematics and Its Applications, for instance). A regular series of International Conferences on the Teaching of Mathematical Modelling was started in 1983 and continues on a biennial basis.

2.2 Early expositions of the methodology of modelling

As a consequence of the increasing appreciation of the importance of mathematical modelling, initiatives have been taken to establish, in undergraduate courses, effective ways of teaching (or encouraging the learning of) the skills of modelling. A *sine qua non* of effective teaching is that teachers have an appropriate understanding of what it is that they are trying to teach. Hence part of the initiative in the teaching of mathematical modelling has been the attempt to establish a more precise understanding of what mathematical modelling is and what characterises an effective modeller. Various models of the modelling process, or methodologies of modelling, have been proposed and a line of development may be discerned.

Drawing on the work of Pollack (1959), Klamkin (1971) proposed a five stage model of the problem solving process, the stages being
1. Recognition
2. Formulation
3. Solution
4. Computation
5. Explanation

Klamkin sees the modelling, or in his terms model building, as taking place in stage two of this process though more recent expositions of modelling would probably argue that all of Klamkin's five stages are aspects of mathematical modelling. More importantly for the present discussion, however, it should be noted that Klamkin described modelling or problem solving as an essentially linear process starting at stage one and proceeding to stage five at which point the problem is solved. On the way each stage in the process is undertaken and completed before embarking on the next stage.

Discussing the role of applied mathematicians and how they should be educated, Lin (1967, 1976) described what he characterised as the creative activities of applied mathematicians. He wrote

Applied mathematicians are often found to be engaged in the following efforts: (a) the formulation of scientific concepts and problems in mathematical terms, (b) the solution of the resultant mathematical problems, and (c) the discussion, interpretation and the evaluation of the results of their analysis including the making of specific predictions.

These activities are, of course, a prescription for mathematical modelling. Lin has introduced the concept of evaluation of the results of the modelling process to the final stage of his description. In a subsequent paper, Lin (1978), he adds another important concept to the model of mathematical modelling, the idea that often there is not a single best or most appropriate model of a physical problem, but rather a range of possible models some of which may be useful for some purposes and others for other purposes. Each possible model may contribute some partial understanding of the physical phenomenon and the modeller must choose the model or models used to meet the requirements of the problem he or she faces.

Woods (1969) provided a description of the process of applying mathematics to physical problems which is very similar to Lin's.

There are three distinct phases in the application of mathematics to natural phenomena, viz.
  (i) the formulation of an idealised model in mathematical language,
  (ii) the solution of the mathematical problem and the deduction of results capable of experimental verification, and
  (iii) the comparison of observation and theory and the evaluation of the validity of the model.
The last step may lead to improvements in the model and the steps are repeated.

There is, here in Woods' description, a stronger implication than in Lin's that the modelling process should be viewed as an iterative one with evaluation of the results of the modelling in the light of physically observable reality leading to modifications of the model and repetition of the stages of the process. Woods also identified a variant of this process in which several models may be competitively evaluated to determine which most closely reproduces the features of the physical problem being modelled – a form of parameter identification problem.

In a discussion of the state of applied mathematics education in UK schools and universities, Ford and Hall (1970) suggested that applied mathematics could be revitalised and set on a firmer foundation than its traditional basis in mechanics if mathematical modelling were accepted as the underlying unifying theme of applied mathematics. They explored some of the implications of such a philosophy and suggested ways in which it could be implemented. Within this discussion they introduced a model of the modelling process which is very similar to the three stage model put forward by Lin and by Woods.

2.3 The development of iterating methodologies

The first methodologies of mathematical modelling, introduced in the previous section, all share the property of being essentially linear and straight through. Subsequent developments resulted in the proposal of a series of models of the modelling process which shared a somewhat different property, the use of some type of loop. These methodologies are still basically linear, in that they consist of a series of stages following on one after the other, but they also have at least one point at which, in a manner similar to a flow chart of a computer program, a branching progression is possible instead of a direct progression to the next step in the linear sequence. Such branching results in a return to some previous stage of the process and the subsequent repetition of some stages.

Hall (1972), building on the work of Ford and Hall (1970), described a model of mathematical modelling containing a path returning from the final validation step to the initial recognition of the real problem and formulation of a model step. This path could be taken if the validation of the model, that is the comparison of the output or predictions of the model system with the equivalent quantity in the system being modelled, revealed inadequacies in the model as a description of the real system which were sufficiently detrimental to the purposes of the investigation being undertaken. A more refined version of the model or even a completely new and different model could then be formulated and the stages of mathematical solution, re-interpretation into real world terms, comparison and validation repeated. Hall commented that

The basis of modelling is the scientific method except that the emphasis is on finding a mathematical form for the scientific theory. The process starts from some given empirical situation which challenges us to explain its obvious regularities or discover its hidden laws.

Bajpai *et al.* (1975) discussed, in some detail, the state of mathematics teaching in engineering degree courses in the UK. Their discussion included the questions of who should do such teaching, what should be taught and to whom, when and how mathematics should be taught to such students, what modes of assessment should be used and why mathematics should be taught to engineers. Amongst the aims and objectives for mathematics teaching in such service courses they included

Students should appreciate the concept of a mathematical model and the methods of obtaining solutions to such a model. They should be aware that the formulation of such a model is an integral part of solving an engineering problem.

They used a flow block diagram to illustrate their conception of the methodology of mathematical modelling. Their diagram is reproduced in figure 2.1.

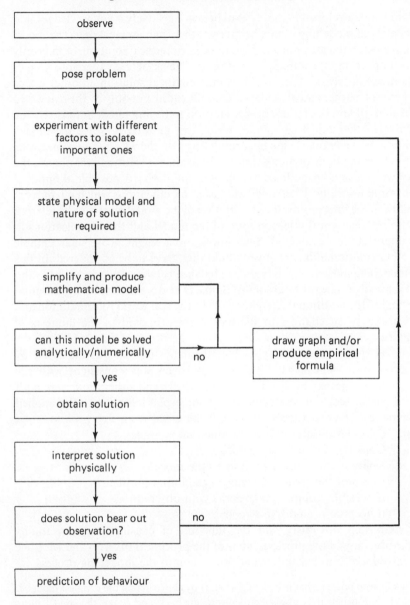

Fig. 2.1. The modelling methodology of Bajpai *et al.* (1975).

The outer return loop from the block 'does solution bear out observation?' to the block 'experiment with different factors . . .' is essentially similar to the branch from validation to formulation suggested by Hall (1972). The inner loop signifies the beginnings of an understanding of a more complex connection between blocks or stages of the modelling process which will be explored and developed in more detail below. Effectively it acknowledges that mathematical modelling is a real activity rather than an ideal one. The model initially postulated may not admit analytic or numerical solution, or may only do so at unacceptable or inappropriate cost. The model may therefore need to be modified purely in order to aid its mathematical solution, in other words the choice of mathematical model used is influenced, not only by the real world problem itself and the nature of the question to be answered, but also by the available mathematical techniques which can be used to solve the model. Initially a modeller may begin by formulating a model which is, in some sense, ideal but, in the light of an assessment of available solution techniques, the model will perhaps be modified to take account of what is possible as opposed to what might ideally be desirable. However, modifications which might seem desirable in the interests of efficient solution of the model must also be referred back to the real world interpretation of the model and their implications in that interpretation evaluated. Such implications might prove unacceptable and other solution aiding changes would then have to be considered. In this way modelling must always be seen as a balance between what is desirable and what is possible. The methodology of Bajpai *et al.* does not explicitly recognise or formalise the additional reference back to the real world interpretation and implications when changes to the model are proposed in the interests of solubility.

A novel treatment of an applied mathematics course was described by d'Inverno and McLone (1977). Their approach was consistent with the philosophy suggested by Ford and Hall (1970) and others and was built around the concept of modelling as an underpinning activity for applied mathematics. Their methodology of modelling was a six step process.

(i)   Select essentials, noting presumed inessentials.
(ii)   Construct an idealised model.
(iii) Apply mathematical analysis to the model.
(iv)   Interpret the results in terms of the initial problem.
(v)   If possible, compare the results with observation.
(vi)   If necessary, modify the model.

The sixth step obviously had the function of causing a return to, by implication, any of the previous steps of the process. d'Inverno and McLone also introduced another important principle into the modelling process.

Another crucial point is that in practice we start by constructing the simplest model. If this is inappropriate it is always possible to go back and make the model more sophisticated, usually by dropping or altering one or more of the assumptions.

McDonald (1977) reported his experience of introducing a course in mathematical modelling into undergraduate degree courses. Drawing on the work of Churchman *et al.* (1957), he defined a methodology of modelling which used a six stage linear progression similar to that of d'Inverno and McLone. McDonald's description did not specifically mention any iterative procedure but Churchman *et al.* are clear that their original work, a systems

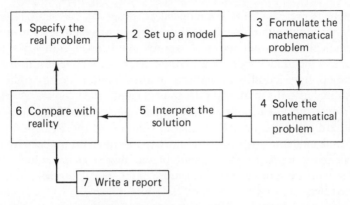

Fig. 2.2. The modelling methodology of Penrose (1978).

modelling methodology, is intended to include the possibility of iteration through the stages of the procedure.

A model of the modelling process presented by Penrose (1978) is divided into six stages in a circular progression. His representation of it is reproduced in figure 2.2. Evidently there is an implication that initial entry is through stage one and exit is via stages six and seven. This methodology has obvious similarities to the procedures described by Hall, d'Inverno and McLone and McDonald. Describing the model, Penrose wrote

> The boxes are numbered, and joined by arrows, to indicate the normal direction of travel from one to another, but it is usual to return to some of the boxes several times as one's ideas develop. In particular, it is frequently necessary to return to Box 1 after reaching Box 6, and to go round the 'loop' a second or even a third time if the first attempt does not produce a satisfactory solution of the original problem. One can regard mathematical modelling as an iterative process, starting from a rather crude model and gradually improving it until finally it is good enough to solve the original problem. This is a much more effective way of tackling modelling problems than trying to get everything right the first time round.

Penrose thus reinforced the point made by d'Inverno and McLone about starting with a simple model and progressively developing more complex versions of it during successive iterations through the modelling process. Such a procedure is obviously only possible with an iterative methodology. Berry and O'Shea (1985), drawing on the work of Penrose and others, also took up the theme of the importance of the iterative nature of the modelling process. The Open University course, MST204, to which Penrose, Berry and O'Shea have all contributed, includes a modelling project in which a percentage of the credit is given for the student's success in iterating through the modelling activities and modifying the model. There is also a hint, in Penrose's description of figure 2.2, of rather more complex and involved feedback loops but these are not represented in the figure.

An example of the development of a model by successive iteration and refinement is presented, for pedagogical purposes, by Clements (1989). An undeniably crude initial model was chosen which was then developed, solved mathematically and validated in an approximate and overall sense. Several iterations of the modelling loop were then made. At each return to the initial formulation stage additional features of the situation being studied were incorporated into the model and the solution and validation stages repeated. This process continued until an acceptable model was reached.

Burkhardt (1977) used a flow diagram similar to the Penrose model to describe the modelling process although he elaborated his model with further inner feedback loops in the manner of the Bajpai *et al.* model. Further discussion and development of this feature of the modelling methodologies is presented below.

Penrose's model also exhibits another major feature of interest, the vertical division of the diagram. The activities contained in the boxes in the left hand column, boxes one and six, are ones which he described as taking place in the real world. Those in the boxes in the right hand column, boxes three and four, were described as being based in the abstract world of mathematics. The central column then contained activities which assume an increased import-ance in that they are the bridges between the real world, where the original problem exists, and the abstract world of mathematics, where the mathema-tical model of the problem is manipulated and solved. This serves to emphasise that the translation of a real problem into a mathematical model and the subsequent interpretation of the mathematical results of the model into their implications for the real problem are crucial activities for the mathematician who hopes to take his or her place as a useful practitioner of mathematics in science, engineering, business and other places where mathematics may be a tool for understanding and changing reality.

The theme of the division of mathematical modelling into those activities which are based in the real world and those which are abstract mathematical activities also appears in the descriptions of modelling given by Burkhardt (1977), Burghes and Read (1978), Clements (1982b) and, though less centrally, in McLone (1976). Burkhardt shows the division into real world and abstract activities very strongly in a diagram in his paper and emphasises that the translation and interpretation activities are essential features of mathematical modelling and ones which have been particularly neglected in mathematics education in the past. A similar diagram appears in Burghes and Read's paper, which appears as a statement of editorial policy in the first issue of the *Journal of Mathematical Modelling for Teachers*. The importance which they place on the bridging skills of translation and interpretation is shown by their statement

Emphasis will be placed on the *setting up* of the model, the *interpretation* of the solution and the *validation* of the model.

The methodology of modelling suggested in McLone (1976) has much in common with the other methodologies reviewed in this section and is evidently related to that of d'Inverno and McLone (1977). It differs in one interesting respect from most of the others; the activity of validation (or comparison with reality) is shown with multiple return paths to nearly all previous stages of the process. McLone suggested thereby that validation is a continuing activity throughout the modelling cycle and that, at each stage, some validation of the partially complete model should be undertaken. In other words, validation is not left until near the end of the modelling cycle, at which point it might indicate the necessity for a repeat of the whole cycle, but takes place during each stage and can lead to partial iterations through various subsets of stages of the complete cycle. Validation is seen as a form of regulatory or controlling activity affecting the whole modelling process. This departs quite strongly from the conception of mathematical modelling as a linear sequence of stages with an optional feedback loop around the whole process, and leads towards the more advanced perception of the methodology of modelling which will be proposed below.

2.4 Methodologies of modelling and the scientific method

The methodologies of modelling which have been described are strongly reminiscent of the scientific method. The linear sequence of activities which make up the scientific method is very similar to most of the sequences which have been proposed as modelling methodologies, viz. observation of the real world, formulation of a hypothesis (or model) to account for those observations, use of the hypothesis to predict further behaviours which the real world should exhibit (or solution of the model and interpretation of the solution into its implications for the real world), validation or refutation of the hypothesis (or model) by comparison of those predictions with the observed universe and possibly revision of the hypothesis (model) in the light of this comparison. This connection is not entirely surprising when it is remembered that most of the traditional practitioners of mathematical modelling have their roots in the mathematical and physical sciences. The more recent users of mathematical models are largely drawn from biological, medical and management areas and even management science may be seen as owing a considerable debt to the scientific method. As already noted, Hall (1972) made the connection between his model of modelling and the scientific method explicitly and Dym and Ivy (1980), introducing their conception of mathematical modelling, say

It is also appropriate to note here that the philosophical idea that we are expounding is not original with us. In fact, it is a hallmark of the scientific method. The underlying philosophical structure has been discussed by philosophers of science, by mathematicians and physicists, by people interested in modelling, such as operations researchers and mechanicians, and by many other practitioners. What we are discussing is really a part of the scientific method.

Having noted that the scientific method is a common strand in the development of methodologies of modelling, it is appropriate to ask if adherence to the scientific method, at least in its simplest form, is entirely inevitable. Before an attempt is made to answer that question it is interesting to note that there is a school of thought in the philosophy of science which holds that the scientific method is not even adequate to explain the process whereby all advances in scientific knowledge are achieved. Kuhn (1962), for instance, identified two mechanisms which lead to advances in the sciences, only one of which matches the classical pattern of the scientific method. It will be argued below that the scientific method is, ultimately, too limiting a basis for the full development of a methodology of modelling and that our understanding of the process must transcend its roots in that method. Unless the inappropriateness, for some modelling purposes, of methodologies based purely in the scientific method is recognised, there is the possibility that profitable lines of development will remain unrecognised and unexplored. It will be argued instead that the systems methodology developed by Checkland (1981) offers insights which can be used in developing a more advanced methodology of mathematical modelling. That methodology is intended to model more precisely the actions and thought patterns of experienced practitioners of mathematical modelling and so give greater insight into the process of modelling.

2.5 The systems movement and Checkland's methodology

The development of systems as an academic and practical discipline has taken place almost entirely within the last forty years. One of the fundamental motivations for the development of systems theory was the recognition that, in a wide variety of scientific, engineering and biological contexts, entities exhibited properties which only became apparent at a certain level of complexity. Such properties defied analysis by the traditional reductionist techniques employed by the scientific method. Instead, the existence of these properties led to the development of a new conceptual framework within which the features of organised complexity could be analysed, classified and discussed. This framework was systems theory, described in its most general form by von Bertalanffy (1973).

The novel feature of the systems approach was that it enabled a holistic study of complex and sophisticated entities and, particularly, the mutual interactions of such entities in hierarchical arrangements and structures – that is the behaviour of systems of systems. Because of its holistic approach, systems theory allows the study of systems composed of entities which are, to a greater or lesser extent, not entirely well understood individually. This is an essential difference from, and in strong contrast to, reductionist methods. The latter would require that complex entities be broken down into simpler and simpler component parts until each component is simple enough to be understood individually, and then assert that, once every component is understood, knowledge of the whole entity is complete. The systems approach denies this and allows the possibility that there are properties of the system as a whole which cannot be understood and elucidated from a study of the component parts of the system but only by a holistic study of the complete system.

To complement this different philosophical viewpoint new methodologies for the organisation of intellectual enquiry were required. Such methodologies need not be seen as in conflict with the scientific approach, but rather should be understood as complementing it and illuminating different facets of a complex and ever-changing world. Indeed these methodologies owe a debt to the scientific method, drawing on some aspects of that method in their development. The debt is acknowledged by Checkland (1981) when he wrote

Hence there is an incentive to examine alternative paradigms to those of natural science, while continuing to build on the scientific bedrock: rationality applied to the findings of experience.

It is not necessary or appropriate to describe the development of systems methodologies as a whole here. It is sufficient to say that one systems methodology, the 'soft systems methodology' of Checkland, has features which may, with benefit, be translated and grafted onto the concensus methodology of mathematical modelling described so far. Such a borrowing greatly enriches the understanding of the process of mathematical modelling provided by the methodology. To achieve this purpose it will not be necessary to describe, analyse and criticise the Checkland methodology in any great detail, but only to delve into those aspects of it which have, in the author's opinion, importance for the future development of methodologies of mathematical modelling.

Systems may be classified into natural (eg biological organisms), designed physical (eg most engineering hardware), designed abstract (eg a stock management system) and human activity systems (any set of appropriately chosen purposeful human activities). Particular systems may not fit neatly into such a categorisation but show aspects of more than one category. Systems theorists also often classify systems as soft or hard systems. These are not discrete categories but rather two opposing ends of a spectrum of variation. Hard systems are essentially those which have an embodiment which is independent of the observer and are thus more susceptible to analysis with traditional mathematical style tools and techniques. Soft systems are those which are ill-structured and where a concensus view of the problem may be more difficult or even impossible to achieve. Human activity systems more often fall at this end of the spectrum. Checkland's methodology is intended to allow meaningful analysis to be made of soft systems problems, though it is applicable to hard problems as well.

It is sufficient for the development of this argument to describe the similarities of Checkland's methodology to that of mathematical modelling. Like modelling, the Checkland methodology consists of a collection of activities or stages. Some of these activities are located in the real world and others are located in an abstract world, not, in this case, the world of mathematics but the world of abstract systems concepts and thought. In these details then the Checkland methodology parallels the concensus we have of the methodology of mathematical modelling. The Checkland methodology has, however, a complex and sophisticated mechanism for linking the stages of activities together and guiding the progression of the systems analyst from phase to phase of the study. Checkland, describing this progression, wrote

. . . although the methodology is most easily described as a sequence of phases, it is not necessary to move from phase 1 to phase 7: what is important is the content of the individual phases and the relationship between them. With that pattern established, the good systems thinker will use them in any order, will iterate frequently, and may well work simultaneously on more than one phase.

The most important idea to grasp from this is the release of the systems analyst from the constraint of a linear progression from the first activity to the final activity of the methodology. The linkage between phases is more complex than that, it is not limited to simple progression from one phase to another with possible feedback or iteration of some phases but rather consists of a series of forward and backward movements between simultaneously active phases of the study with the possibility, in any given change, of moving around one or more intervening phases. This creates a much richer set of possibilities and a more complex structure for the methodology.

It is this much more complex and richer pattern of interconnections and interactions between the stages of the Checkland systems methodology which, it will be suggested in the next section, can with benefit be grafted onto the methodology of mathematical modelling which has so far been built up in this chapter. The result is a richer and more capable methodology of modelling and one which more truly captures and describes the spirit of the modelling activities of experienced mathematicians.

2.6 A complex linkage methodology of mathematical modelling

It has been increasingly the author's experience that the established methodologies of mathematical modelling, with their scientific method based linear progression from step to step of the modelling process and their basically single looping structures, are insufficiently rich to describe completely the ways in which experienced mathematical modellers work. Clements and Clements (1978) commented

The model and analysis chosen will be decided by both mathematical and non-mathematical factors, eg the type and accuracy of the answer needed, the time and manpower available to devote to the problem, and the computer power and financial resources available.

That comment was intended to apply to mathematical modelling and problem solving in an industrial or commercial environment, but the point may be extended to embrace all mathematical modelling. The experienced modeller's activities and thought processes whilst working on any given stage of the modelling cycle are conditioned by knowledge of the implications, possibilities and limitations imposed by considerations arising in other preceding and succeeding phases of the cycle.

If the modeller commenced on a strictly linear path through the modelling cycle it is possible that he or she would arrive, for instance, at the stage of interpretation of the mathematical answer back into real world terms only to find that the model adopted had been far too complex, or addressed an aspect of the problem which was irrelevant or peripheral to the real world concern, bearing in mind the nature of the solution which was desired or the fundamental objective of the study. This could potentially result in considerable wasted effort. A much more appropriate mode of working, and one which experienced modellers do indeed exhibit, would have been, during the model formulation stage, to be already considering the mathematical analysis likely to be possible on such a model, the results which would be obtained thereby and their possible interpretation. Notice that this would involve looking forward not just to the next step in the modelling cycle but even further. In the light of such forward considerations, work on the current phase of the cycle might be modified. Again, suppose that the mathematical solution of a model were undertaken in isolation from the preceding and succeeding activities. Under such circumstances it is possible that a solution method might be used which was quite inappropriate in view of the nature of the real world solution required or perhaps the nature of the approximations which had been made in the formulation of the model or the accuracy of the data found to be available during the preliminary analysis of the real world problem. Perhaps, if the solution method adopted were a numerical one, numerical difficulties or instabilities might occur. To suppose that these must be met and overcome as a problem purely in numerical analysis is to ignore the possibility that the numerical difficulties might be indicative of faulty formulation of the model or of the critical nature of some feature of the real problem which had been assumed to be unimportant. Again this suggests that the optimum mode of working will involve frequent reference forwards and backwards between phases of the modelling cycle. Suppose the modeller were to insist on completing the preliminary assessment of the real world problem before even beginning to consider what models might be created, what analysis might be undertaken and how that might translate into real world solutions. It is probable that effort would be wasted investigating aspects of the problem which need not or could not be included in the model, or which were irrelevant in the light of other approximations being made.

All of these examples serve to illustrate the point that the optimum mode of working on a mathematical model is not to follow the sequential modelling cycle from analysis of real world problem, through creation of mathematical model, solution of the model, interpretation of that solution into real world terms and finally evaluation of the solution followed possibly by a return to the analysis, some modification of the model and a repetition of the subsequent phases of the cycle. The better mode of working might be thought of as a parallel attack on all phases of the modelling cycle simultaneously but with effort focussed at any given time more on some particular phase or phases than on the remainder. The focus of attention of the modeller will then move through something approximating the basic linear sequential cycle of the modelling process but not in a slavishly constrained way. In this way considerations of practicality and indications of effective modes of attack may carry across from any phase of the cycle to any other phase in a continuous and natural manner. Alternatively, the pattern may be thought of as a basic linear sequence of activities to be traversed, possibly several times, during a study, superimposed upon which is a complex pattern of interconnections and linking paths which will be repeatedly traversed during each cycle of the main loop. Finally this leads to a model of the methodology of mathematical modelling which may be summarised diagrammatically as in figure 2.3.

The methodology illustrated in figure 2.3 distinguishes between the concrete entities or products involved in mathematical modelling and the activities or processes which connect these. Thus the concrete entities are the original real problem, the mathematical model which is created from or recognised within that problem, the mathematical solution which arises from computation or analysis of the mathematical model, the real solution which results from the interpretation of the mathematical solution into the terms of the original problem and the proposed actions which result from the evaluation of the real solution. These concrete entities, as well as being connected by the primary paths of the modelling process, are associated with each other through

Fig. 2.3. A richly linked model of mathematical modelling.

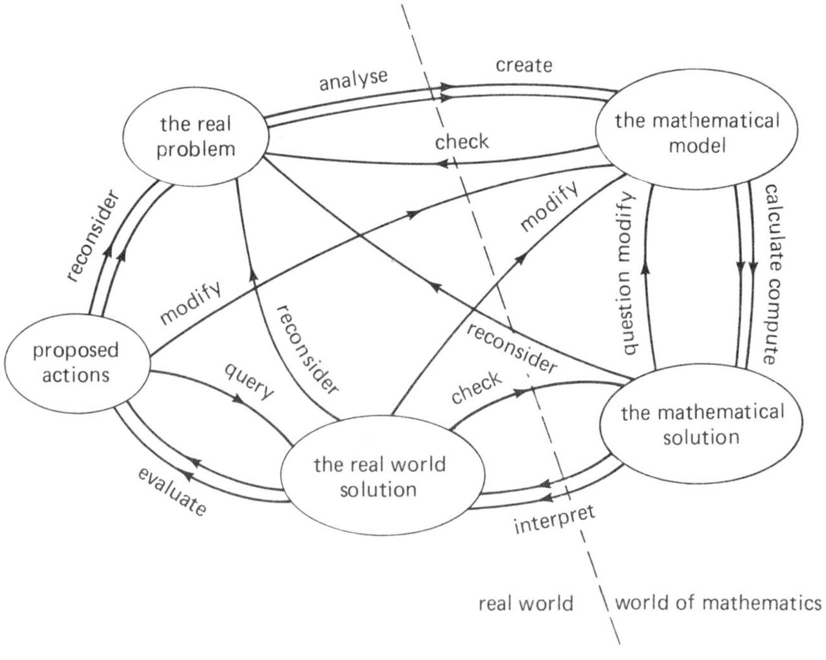

subsidiary paths. The repeated traversal of these subsidiary paths results in many partial loops through various subsets of the stages of the modelling process.

2.7 What use is a methodology anyway?

Chapter one suggested two distinct categories of people who might benefit from knowledge of a paradigm of the modelling process. Firstly, lecturers and teachers should find that some study of the methodology of mathematical modelling is useful both in clarifying what it is that they wish to teach (or wish to help their students to learn) and in helping them to recognise the extent to which their students are indeed learning the skills of modelling.

Secondly, it was suggested that, after an initial exposure to modelling as a practical activity, students might benefit from some theoretical reflection on what they had been doing. Such theoretical study helps to place the practical skills in context and provide a mental framework in which to order the otherwise unstructured fragments of knowledge and technique which they have acquired. This does not imply, however, that it is necessarily appropriate to introduce students to a very full and complex model of the modelling process immediately after a first practical course in modelling. On the contrary, there is every reason to suppose that learning is best achieved by a process of progressive revelation. That would suggest that the first methodology of modelling which is given to students might well be a simple one, more akin to the straight through, linear models which have been discussed in this chapter. Once a relatively simple model has been understood and internalised, and some more practical experience has been gained using this simple model to guide the process, students may then be receptive to the subtler and more complex ideas put forward here. Taught in this way the twin developments of intellectual knowledge concerning the modelling process and practical skill in developing models are interlinked and each supports the other.

2.8 Conclusion

This chapter has attempted to present a background to the current state of mathematical modelling as a teaching activity. Firstly we described the development, within the academic community, of the twin realisations that students needed specific help in learning the skills of modelling and that means of giving such help in a fairly directed way could be devised. Next the development of models of the mathematical modelling process was traced. A new way of characterising the linkage and progression between stages of the modelling process, derived from the work of Checkland on systems methodologies and first proposed in Clements (1982c), was then suggested and, finally, some comments on the place of the study of methodology in the teaching of mathematical modelling were made.

3

Mathematical modelling in practice

This chapter attempts to identify and illustrate some of the practical skills needed by an effective mathematical modeller. It does this partly by direct comment on the modelling process and partly through the medium of a number of examples of the development of simple mathematical models. It should be apparent that no description of the processes of mathematical modelling can ever hope to be absolutely complete or to encapsulate the 'last word' on the subject. In that spirit then, this chapter draws on experience accumulated over a number of years by the author both in his capacity as a mathematician working in an engineering environment using his mathematical knowledge in the solution of engineering problems and in his capacity as a teacher of mathematics and mathematical modelling.

3.1 Understanding the problem

The first general comment to be made is that very little progress can be made in modelling if the modeller does not fully understand the system or situation which he or she is trying to model. (Before going any further let us, for the sake of brevity and clarity, acknowledge that the subject of a mathematical study may be any of a wide spectrum of entities ranging from concrete physical mechanisms through to relatively abstract systems of human activities and therefore agree to denote all of these possibilities by the single term 'subject'.) The first stage of modelling should be the collection of data or experience about the subject to be modelled. This study will usually be more qualitative than quantitative. Such a study will entail, perhaps, the examination of some design drawings of a mechanism leading to an understanding of the basic mechanical or physical processes taking place in the operation of the mechanism, or talking to and questioning those involved in a human activity system in such a way as to gain an understanding of the basic interactions of those involved. Even at this stage, though, an experienced modeller will be working through some preliminary ideas in a mathematical way and looking ahead to the future stages of the modelling process. Such preliminary mathematical working and looking ahead often helps to clarify the modeller's physical understanding and to highlight possible misconceptions which may otherwise build up in the modeller's mind.

It is worth mentioning here one important reason for identifying and eliminating misconceptions at as early a stage as possible. It is a very common human trait to rapidly become extremely committed to one's conceptions of a problem. It is, as a result, common to find engineers and mathematicians clinging to unlikely or demonstrably wrong ideas about the mode of operation of a modelling subject. Such erroneous ideas need to be identified and eliminated at as early a stage as possible to minimise the commitment they have engendered and make their rejection and replacement by other concepts more readily achieved.

The problem described in O'Carroll (1981), the modelling of the operation of a hydraulic buffer, illustrates rather well the problems which may be engendered by a faulty understanding of the nature of the problem and the way in which the mathematical working through of even preliminary ideas about a problem can help to illuminate physical understanding of the modelling subject. Early in the chapter O'Carroll describes the misconceptions about the principle of operation which many students, some quite experienced, attempting to model the buffer have demonstrated. The author has also used this problem with second year undergraduates and found that many, if not most, of them have shared the misconceptions of O'Carroll's students. A typical buffer of this type is illustrated in figure 3.1 in its fully extended and its fully compressed states.

Most students immediately conclude that the operating principle is that the gas in the buffer is compressed providing a cushioning spring effect. A very brief mathematical exploration of this idea will soon suggest that such a mechanism cannot possibly provide a sensible explanation of the operation of the buffer. The argument for this conclusion would proceed roughly thus. From the diagram it is evident that the compression ratio of the gas between fully expanded and fully compressed states is roughly 4:1. The travel of the buffer is roughly 0.1 m and, on the basis of the scale drawing, the radius of the gas container may be estimated at 0.05 m. The energy required to compress the gas to its final volume is

$$E = -A \int_{x_0}^{x_f} P \, dx$$

where A is the internal cross-sectional area of the gas vessel, P is the pressure of the gas and x_0 and x_f are the initial and final lengths of the gas vessel. Using the adiabatic gas law

$$PV^{\gamma} = \text{const},$$

the relationship

$$V = Ax$$

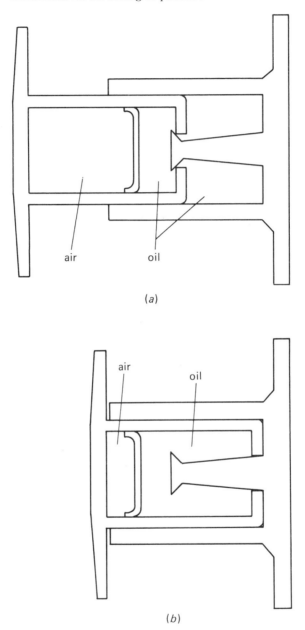

Fig. 3.1. Hydraulic buffer extended and compressed. (*a*) extended configuration (*b*) fully compressed configuration.

between the volume of the gas vessel, V, and A and x and the 4:1 estimated compression ratio between the extended and compressed states of the buffer, the energy needed may be estimated as

$$E = \frac{(1 - 4^{\gamma - 1})}{(1 - \gamma)} A P_0 x_0 = 1.853 \, A P_0 x_0$$

where P_0 is the initial pressure of the gas. Using the estimate already made of the radius of the gas vessel cross section we may finally estimate that the energy needed to compress the gas is of the order of 0.15 kJ if the initial pressure P_0 is atmospheric, and this energy scales linearly with initial pressure P_0. Typically, in a railway application, a pair of such buffers might be called upon to halt a truck of 50 tonnes from a speed of $2 \, \text{m s}^{-1}$. The energy of such a truck is of the order of 100 kJ. Evidently either the buffer is initially charged with compressed air at around 330 atmospheres or the energy must be dissipated elsewhere than in compressing the air in the buffer!

Despite such an obvious discrepancy it is surprising how many student groups cling to their first idea that the principle of operation is that of an air spring. Most seem to persuade themselves that the problem lies not in a misconception about the physical understanding of the problem but in some error caused by their arithmetical and manipulative ability or lack of it.

The argument outlined above basically involves hypothesising about the underlying physical mechanism of the subject system, looking ahead to the implications of this hypothesis in the next stage or two of the modelling process, seeing that a fundamental inconsistency would arise and returning to make another attempt to understand the physics of the problem. This example therefore illustrates rather well the comments which were made in chapter two about the complexity of the interactions between the various stages of the modelling process and the way in which experienced modellers actually work with an eye to the implications of what they are doing for several phases of the modelling process simultaneously.

Another point which is illustrated by this example is the prime importance of the ability to make sensible assumptions, approximations and estimates. In the course of the argument about the energy required to compress the gas several parameters (compression ratio of the gas and stroke of the buffer between fully extended and full compressed positions, and radius of the gas chamber) of the buffer were estimated from the drawing of the buffer. None of these estimates are particularly accurate. In view of the size of the mismatch found between the energy needed to compress the air in the buffer and the energy to be dissipated in the moving truck the errors of estimation are insignificant. Nonetheless it is typically the case that students and other inexperienced modellers will concentrate considerable effort on providing the most accurate possible estimates of these quantities. The author has sometimes found student groups virtually refusing to proceed further until they have been provided with accurate data! The experienced modeller, on the other hand, will proceed using crude estimates but keeping in mind the low quality of the data used. If the result had been to find that the energy of compression of the gas and the energy of the truck were even within one order of magnitude of each other then data of better quality might have been called for. In the event there is little point in refining the estimates of any of the physical parameters. The author has frequently characterised the quality which is required of the modeller in this context as 'mathematical discretion'.

Returning to the main point of this section, another way in which some student groups on the author's courses have gained valuable insight into the physics of modelling subjects, particularly where the subject is a physical mechanism, is through working models. Such models do not usually need to be highly accurate as regards scale et cetera nor very sophisticated as regards construction methods or materials. Meccano and other commercial component based constructor outfits have been used and one of the best and most revealing models the author has been shown was constructed from firelighting spills and dressmaking pins! The prime purpose of most of the models which students have constructed to the author's knowledge has been to illuminate their understanding of the operation of mechanisms and emphatically not for the purpose of making measurements or carrying out real experimental work. Crude and unsophisticated models are adequate and even ideal for these purposes. They would, obviously, not be adequate for more exacting purposes.

3.2 Iterative refinement of models

The second general comment to be made on the modelling process is that it is often easier to refine an existing model than to create a new one from scratch. By the same token it is usually easier to create a complex model by a process of successive refinements, initially creating a simple and quite possibly inadequate model of the subject then refining that simple model to take account of increasingly complicated and sophisticated features of the subject until the final model is usable for the purpose in hand. The author has observed that students rarely adopt this type of approach spontaneously. More commonly they identify a very comprehensive list of the features of the real world subject which they feel would need to be accounted for in their model and then set about creating, in one stage, a complete model incorporating all these effects. The result of such an approach is usually confusion! It is almost always more practical to start by choosing one or two of the effects thought to be most important or dominant and creating a model which accounts for these effects only. When such a model has been created and evaluated it is much easier to add the effects of the lesser considerations to this backbone model, testing and evaluating each of the modifications one by one, until a complete model has been constructed.

An additional advantage of this type of successive refinement approach is that, at each stage, the model created so far can be examined and its implications for the real subject being modelled worked out. This process has three useful effects. Firstly, it reveals any inconsistencies between the model and reality. Such inconsistencies usually indicate arithmetic, algebraic or other mathematical errors in the formulation of the model or misconceptions about the real subject which have been built into the model. Either of these problems can be promptly corrected. Secondly, inadequacies in the model as a description of reality will be revealed and examination of these will focus attention on those aspects of the model which most need refinement. Thirdly, in the process of comparison with reality, new lines of mathematical approach are often suggested by the physical or abstract reality of the problem. The author has, in the past, often characterised this process as 'listening both to what the mathematical model is trying to tell us about the reality of the modelling subject and to what the real modelling subject is trying to tell us about the mathematical model'. The whole process is one of mutually supportive activity in both mathematical modelling and understanding of the real world.

To illustrate this type of model construction we will study the problem of temporary traffic lights. This problem was originally prompted by the author's observation of a set of temporary traffic lights encountered on a major road during a car journey. The road normally carried a single lane of traffic in each direction. Road works had resulted in this being reduced to a single lane in one direction or the other, the current direction being controlled by two linked sets of temporary traffic lights, one at either end of the single lane section. The single lane section was fairly long, about 600 m, but the lights were set in such a way that only five or six cars were able to pass through the section in each cycle of the lights. This seemed to be somewhat inefficient since large parts of the road were unoccupied for much of the time. The question posed in the mind of the author was 'Is it possible to create a mathematical model of the operation of the traffic in such a section of the road and to use it to predict an optimum cycle time for the lights given the length of the single lane section, the maximum safe speed of the traffic through the section and any other relevant parameters?'.

As a first attempt at such a model let us suppose that a typical set of temporary traffic lights is arranged as in the schematic diagram of figure 3.2. Let us assume that a cycle of the system consists of the light at A showing green for a time T_g allowing traffic to pass in the direction A to B. Following this both lights show red for a time T_r to allow the single lane section to become clear of traffic. Subsequently, assuming the operation of the lights is symmetric with respect to direction, the light at B shows green for a time T_g, allowing traffic to pass from B to A, and finally both show red again for a time T_r. In the initial model we will simplify things by ignoring the amber phase which usually occurs between the green and the red phase and the red and amber phase occurring between the red and the green phase.

When a light is green the speed and rate at which traffic passes will be a function of the road conditions. If the surface is poor, the carriageway restricted and heavy machinery is operating alongside on the closed section drivers will proceed more slowly and cautiously than if any or all of these

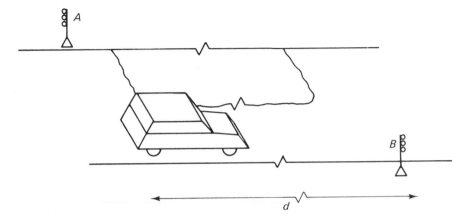

Fig. 3.2. Schematic diagram of temporary traffic lights.

conditions are absent. We will assume that traffic passes in the appropriate direction at a speed $w \, \mathrm{m \, s^{-1}}$. The traffic flux (the rate of passage of vehicles past a fixed point), $q \, \mathrm{veh \, s^{-1}}$, and the velocity of the vehicles are, in road traffic theory, generally assumed to be a function of the traffic density, see for instance Whitham (1974), Dym and Ivy (1980) or Drew (1981). Various models of this relationship have been proposed. We will, for the moment, just note that, through such relationships, we may consider q as a function, $q(w)$, of w. It is apparent that, when a light turns to green, traffic will not immediately pass at speed w, rather there will be an interval in which traffic accelerates to that speed and the effect of this will be to reduce the total traffic passing in any given green phase. This effect will become relatively more pronounced and important as the duration of the green phase is reduced. In keeping with the philosophy of firstly creating a simple model and then subsequently making it more sophisticated and capable, we will ignore this effect in the first instance. Hence the initial model for the number of vehicles passing through the single lane section in one green phase is $q(w)T_g$. During the red phase the last vehicle to enter the single lane section must clear the section so we must have $T_r \geq d/w$. The flux of vehicles in each direction is then

$$\frac{q(w)T_g}{2(T_g + T_r)}.$$

This would be maximised if T_r were made as small as possible. Thus, taking $T_r = d/w$ and putting $T = T_g + T_r$, the maximum possible flux of vehicles is given by

$$\tfrac{1}{2}\left(1 - \frac{d}{wT}\right)q(w). \tag{3.1}$$

This is effectively the capacity of the road whilst obstructed by the single lane section.

Let us now examine the implications of this initial model. The capacity of the road evidently increases as the half period of the traffic light cycle, T, increases and, in the limit as T becomes infinite, becomes $\frac{1}{2}q(w)$. The implication of this is that the capacity of the road, when traffic can only flow in one direction at a time, is always less than a half of the flux of vehicles along the single lane section. The maximum flux is achieved when the cycle time of the lights is very long and the flux decreases as the cycle time of the lights decreases. All these conclusions seem consistent with what we might have expected from the physical situation.

The comparison of the initial model with reality then gives us confidence that we are working along the right lines and that our mathematical activity yields results which are consistent with our understanding of the real modelling subject. It also suggests that, if the expression 'optimum cycle time for the lights' in the problem which we set ourselves before we started is interpreted to mean that the traffic flux along the road should be maximised, the lights should allow traffic to pass in one direction for a very long time before changing to allow traffic to flow in the opposite direction. This would lead to very large delays for some drivers and, instinctively, we feel that this is not allowable. Comparison of the model with reality therefore focusses our attention on another aspect of the problem, the mean delay suffered by vehicles as they pass through the carriageway restriction.

In order to model the delay suffered by individual vehicles let us suppose that the actual flux in either direction is n veh s^{-1}. Obviously n must be less than $\frac{1}{2}(1 - d/wT)q(w)$, otherwise the demand exceeds the capacity of the restricted section and the queue of vehicles waiting to pass the restricted section will grow indefinitely. During the red period of one light $(T + T_r)n$ cars will collect in a queue. When the light turns green the queue will disperse at $q(w)$ veh s^{-1}. We have not, so far, made any assumptions about the regularity with which the n veh s^{-1} arrive. In order to derive an expression for the mean waiting time of vehicles passing through the system we will assume that vehicles arrive in a steady, regular stream. Hence the waiting times of the first, second and subsequent cars are $T + T_r$, $T + T_r - 1/n + 1/q(w)$, $T + T_r - 2/n + 2/q(w)$, ... Let us denote the quantity $1/n - 1/q(w)$, the difference between the inter-arrival interval at the queue and the inter-despatch interval from the queue, as δ seconds. Summing and averaging these over the number of cars in the queue yields an average waiting time of

$$\frac{1}{2}(T + T_r)\left(1 + \frac{n}{q(w)}\right) - \frac{\delta}{2}. \tag{3.2}$$

Examining this expression we see that the mean waiting time increases as n increases and so, observing the restriction that n must be less than the capacity of the restricted section of road, the largest mean wait will occur when n is equal to the capacity of the restricted section. In any case the mean is minimised by taking T_r and T as small as possible. This suggests that the red phase should be as small as possible subject to its restriction $T_r \geq d/w$ and that the half period of the light cycle should be as small as possible.

Thus, the refined model now suggests that, in order to optimise (maximise) the capacity of the restricted road, the cycle time should be as large as possible whilst, in order to optimise (minimise) the waiting time of individual vehicles, the cycle time should be as short as possible. A suitable strategy for determining the cycle time of the lights would be to measure the maximum demand in terms of vehicle flux and set the cycle time to the smallest value at which the capacity of the restricted section, as given by equation 3.1, meets the demand. This is achieved when

$$T = \frac{q(w)}{q(w) - 2n} \frac{d}{w}.$$

The mean waiting time of vehicles is then minimised – any further reduction of T will result in demand exceeding capacity and the queue of vehicles waiting to pass the restricted section increasing without bound.

The alert reader will have noticed, amongst other flaws in our model so far, that the queue was assumed to contain only vehicles arriving during the red period. Because the queue formed during the red period takes a finite time to disperse, the first few vehicles to arrive during the subsequent green period will also be compelled to queue. Starting at the point that a light turns red and assuming that vehicles arrive uniformly at rate n veh s^{-1}, the arrival times at the queue are $1/n, 2/n, \ldots, k/n$ seconds. The equivalent despatch times are $T + T_r + 1/q(w)$, $T + T_r + 2/q(w)$, $\ldots$, $T + T_r + k/q(w)$. Thus the first vehicle which does not have to wait is the k'th where k is the smallest integer for which $k/n > T + T_r + k/q(w)$. Taking k as $(T + T_r)/\delta$ (ignoring the possibility that this may not actually be an integer value) and summing and averaging the waiting times of the vehicles in this longer queue reveals that the mean waiting time is

$$\frac{T + T_r - \delta}{2}. \tag{3.3}$$

Of course, the vehicles which arrive after the queue has cleared but before the lights again turn red will pass without queueing. Taking account of this also the mean wait may be computed as

$$\frac{(T + T_r)(T + T_r - \delta)}{4nT\delta}. \tag{3.4}$$

Considering the implications of both of these new expressions for the real modelling subject we see that neither of them alter the conclusion that the waiting time is minimised when T is minimised subject to the appropriate constraint.

We commented above that, in the first instance, we would ignore the finite time taken by vehicles to accelerate to the speed w m s^{-1} at which vehicles pass through the single carriageway section. We will now refine the model further to take account of this aspect. Let us assume that, when the lights change from red to green, the leading vehicle accelerates at a uniform rate a m s^{-2} until it reaches w m s^{-1}. Using the well known standard formulae for constant acceleration motion we find that the vehicle takes w/a s to reach w m s^{-1} during which time it covers $w^2/2a$ m. Assuming that d, the distance between the two sets of temporary lights, is greater than this, the remainder of the distance, $d - w^2/2a$ m, is covered in $d/w - w/2a$ s. By the time the leading vehicle passes the end of the restricted section the following vehicles will have accelerated in such a way that they will be travelling at w m s^{-1} with flux past the end of the section $q(w)$. Hence the number of vehicles passing in one cycle of the lights is

$$[T - (d/w - w/2a) - w/a]q(w)$$

so the net flux of vehicles is

$$\tfrac{1}{2}\left[1 - \frac{d}{wT} - \frac{w}{2aT}\right]q(w).$$

This gives a slightly lower flux than the previous model. Consideration of the physical reality of the modelling subject shows that this is as we would expect since the effect of the finite acceleration time is to delay the first vehicle through the system during each phase of the lights so reducing the capacity.

We could now go on to consider whether the finite acceleration time affected the argument used to derive the mean waiting time. The purpose of describing this model was, however, not so much for the sake of deriving the model itself as in order to illustrate the method of building mathematical models by iterative refinement. It will therefore be left to you, the readers, to refine this model further if you so wish.

In this section the benefit of an iterative approach to the construction of mathematical models has been urged. In such an approach the initial modelling is carried out by concentrating on the effects of a minimal number of factors chosen from those which are expected to be most significant. Successive iterations then bring into the model further factors of less central importance. The iterative approach allows frequent comparison of the model with the modelling subject. Such comparison ensures that development of the model remains relevant and appropriate and takes maximum advantage of the assistance which consideration of the real subject can give to the modeller.

3.3 Mathematical constraints on models

The third general comment to be made is that it is quite common for the mathematical model chosen for a particular purpose to be motivated and even constrained by considerations which are peripheral to the modelling subject. Comments on the way in which this occurs through forward and backward referencing between stages of the modelling process have already

Fig. 3.3. First model of the human body.

been made in the previous chapter. An example of the sort of constraint which may affect choices made during model construction is the mathematical tractability of the model which will ensue if different choices are made.

As an example of the operation of this type of constraint let us consider the problem of estimating the surface area of a human body. Such a problem might arise in many contexts – for instance in environmental engineering when the heat radiative capacity of the body might be required or in a variety of medical contexts – but let us, for the moment, not be unduly concerned with the reason but merely suppose that we need an estimate of the surface area of a human body.

The author's initial idea about this problem was to divide the body into six basic components, head, body, two arms and two legs, and to model each of these separately. The first model sketched out represented each component as a rectangular parallelepiped as illustrated in figure 3.3. To use such a model to estimate the surface area of any particular human would then require measurements to be made of the length, width and breadth of each component. Given these measurements the surface area of each component (ignoring facets which abut each other) can easily be worked out. Further, the volume of each component could also be computed and the total volume of the body estimated as well as its surface area. Given that, by adjusting the volume of the lungs, a confident human can arrange to float or to sink in a swimming bath, it is reasonable to assume that the specific gravity of the human body overall is approximately unity. Hence the volume of the body and the weight of the human concerned, which is readily determined, should approximately correspond. This provides a valuable check on the results of the modelling – although correspondence of the volume and the weight do not guarantee accuracy of the surface area estimate, a large discrepancy of the two would certainly cast doubt upon it.

The drawback of this first model was that establishing the necessary dimensions was difficult. The human body does not really comprise a collection of rectangular parallelepipeds so estimates of the dimensions of rectangular parallelepipeds of volume equivalent to each component of the body would have to be made by eye. Thus the model is deemed less than perfect not only from the point of view of accuracy of representation of the body but also because of the practical difficulty of making the measurements required to use the model. Hence the desirability of modifying the model was a function of practical considerations as well as modelling ones.

Having established the concept of dividing the body into components and computing both surface area and volume of each component the next stage was to consider other component shapes which might be used. The author's second idea was to use spheres and cylinders as shown in figure 3.4. These components also have well established expressions for both surface area and volume in terms of the diameters and lengths of the various components. Whilst the diameter of an arm or leg is probably no easier to estimate directly than its width and breadth, it is possible to measure the circumference of the limb easily (with a tape measure or a piece of string) and then to derive its radius. Using the same technique to estimate width and breadth required for the previous model would involve an extra assumption (perhaps that the limb is square or has a particular width to breadth ratio). Hence the second model seemed more useful both because it would be easier to make the measurements required to apply the model to any particular individual and because the component shapes seem, intuitively, closer to those of the body's component parts.

Fig. 3.4. Second model of the human body.

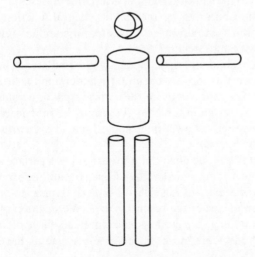

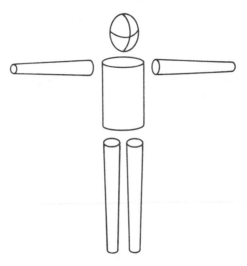

Fig. 3.5. Third model of the human body.

This second model led naturally to a third iteration in which the individual components were modelled by shapes more nearly related to the shapes of the human limbs. Figure 3.5 illustrates a model in which the head is represented by an ellipsoid with three unequal semi-axes, the body by a cylinder of elliptical cross-section, and legs and arms by tapered cylinders of circular cross section. Expressions for volume and surface area of these solids were then sought. Such expressions were found in standard reference works or derived using integration techniques for each of the solids except the spheroid. In that case, whilst the volume is simply expressible in terms of the semi-axis lengths the surface area is not. Using this model would, therefore, require numerical integration techniques to be used to calculate the surface area of the head element. The surface area of the elliptical cylinder is also not representable by a simple algebraic expression but involves the elliptic integral function of the second kind. These functions are, however, tabulated in standard reference works such as Abramowitz and Stegun (1965).

Finally a hybrid of the second and third models was formulated. The head was represented as a sphere, the body as an elliptic cylinder and the arms and legs as tapered cylinders. The expressions for the surface areas and volumes of the various components are given in table 3.1. To prove the practicability of the model the author measured his own limbs and estimated his own surface area. The results are shown in table 3.2.

From these results it may be concluded that the author has a surface area of approximately 1.67 m^2 and a volume of 61.7 litres. Given the author's weight of 67 kg this implies a specific gravity of 1.09. This is sufficiently near to unity to give some measure of confidence in the result for area. We might expect the area estimate to be, if anything, on the low side since the effect of various protruberances on the body (ears, nose, fingers, toes et cetera) have been

Table 3.1. *Expressions for surface area and volume of body components*

	Volume	Surface area
Head sphere	$4\pi r^3/3$	$4\pi r^2$
Body elliptic cylinder	$\pi r_1 r_2 h$	$4h r_1 \mathrm{E}[(r_1^2 - r_2^2)/r_1^2] + 2\pi r_1 r_2$
Arm tapered cylinder	$\pi(r_1^2 + r_1 r_2 + r_2^2)h/3$	$\pi(r_1 + r_2)\sqrt{(h^2 + (r_2 - r_1)^2)} + \pi r_2^2$
Leg tapered cylinder	$\pi(r_1^2 + r_1 r_2 + r_2^2)h/3$	$\pi(r_1 + r_2)\sqrt{(h^2 + (r_2 - r_1)^2)} + \pi r_2^2$

Table 3.2. *Measurements of body components*

	Volume/cm^3	Surface area/cm^2
Head	4046	1232
Body	32233	6440
Arms	6412	3102
Legs	19020	5946
	61729	16720

Head – Three measurements of circumference, 55, 62 and 70 cm averaged to give a mean radius of 9.9 cm
Body – Width 41 cm, circumference 92 cm and length 70 cm
Arms – Circumference at shoulder 34 cm and at wrist 17 cm and length 60 cm
Legs – Circumference at thigh 54 cm and at ankle 24 cm and length 75 cm

ignored in this simple model. If a more accurate estimate were required the model could obviously be refined further to account for these elements.

The point of describing this model here was, however, not to show how a model of the human body could be constructed but to illustrate how constraints of practicability and tractability naturally affect the activity of the mathematical modeller. In this case the model was established in an iterative fashion, at each stage considering the way forward to the solution of the mathematical problem posed by the abstract model and, where appropriate, choosing model representations of the real world or changing the model not just for reasons of accurate representation of the real world subject but also to ease the mathematical solution of the model. In this case the final model settled on a sphere as a representation of the head when, purely from a modelling point of view, an ellipsoid might have been preferred. The compromise was made for the sake of mathematical tractability.

3.4 The importance of choice of notation

The fourth general point to be made is that sometimes the felicitous or deliberate choice of a suitable notation for the expression of the mathematical model has a considerable effect on the tractability of the mathematical problem posed by the model.

Clements (1982b) described a problem of modelling the possible arrangements of the petals of a simple flower shape. In this type of flower (the original subject was a dog rose) a single layer of petals is arranged radially around a central structure with adjoining petals overlapping each other. Each petal may lie above or below its anticlockwise and clockwise neighbours. Thus, imagining we look down on the flower from its face side, each petal may be one of four types, type A, lying over both its neighbours, type B, lying under its anticlockwise and over its clockwise neighbour, type C, lying over its anticlockwise and under its clockwise neighbour, or type D, lying under both its neighbours. Obviously there are constraints on which type of petal may neighbour which type. Proceeding clockwise around the flower, a petal of type A or B must be succeeded by a petal of type B or D and one of type C or D by a petal of type A or C. Abstracting the problem to an n-petalled flower the problem then becomes that of enumerating the possible circular permutations of n objects chosen from a pool of objects of four types subject to constraints of the types which might neighbour each other.

An alternative way of looking at the same problem is to consider the petal junctions. These are of two types, anticlockwise petal over clockwise petal or anticlockwise petal under clockwise. In this case an n-petalled flower has n petal junctions each of which may be one of the two possible types. There are no constraints on which type may follow which type so the problem reduces to the mathematical one of enumerating the possible circular permutations of n objects chosen without constraint from a pool of objects of two types. This is a much simpler way of looking at essentially the same mathematical model of the problem. Interestingly, it should be noted that the problem of circular permutations is considerably more complex than that of linear permutations. The point of describing the problem here is to illustrate how much simpler the problem statement becomes when a certain notation is used to describe it, so it is not appropriate to describe the solution in detail. For the interested reader the solution to the problem may be found in chapter two of Hall (1967).

Another example of the same effect is provided by Stern (1987). In his paper Stern describes a mathematical model used to disentangle a set of relatively complex consanguinity regulations laid down in the rabbinic law of the Jewish faith. The problem stated in prose text seems extremely intractable but the elegant notation described by Stern results in a relatively transparent rendering of the problem into a mathematical form.

The point which this section seeks to illustrate is that similar decisions about choice of notation often have an important effect on the utility and tractability of mathematical models of many kinds. The expenditure of effort and thought on the development or choice of notation is frequently effort well expended.

3.5 Data and sensitivity of conclusions

The fifth general point to be made concerns data and its relation to mathematical models. Students studying mathematics in tertiary education institutions generally expect a certain reciprocal principle to be applied to all the problems which they will meet. This principle may be stated as 'all the data needed is provided and all the data provided is needed'. This principle may be extended further by inclusion of a rider so that it becomes 'not only is all the data needed provided and all the data provided needed but the accuracy to which the data is provided is appropriate to the problem'. In the real world this principle most decidedly does not apply. The major part of some problems consists in collecting the data needed. In others a major activity is distinguishing between the data which are needed and those which are irrelevant.

It is also important to be able to decide or determine the appropriate accuracy to which data are needed. It is often the case that approximate values are readily available whilst obtaining more accurate ones will require considerable expenditure of effort. If, under these circumstances, it can be shown that the approximate ones suffice for the purpose in hand much effort can be saved. An example of such a case arises in the modelling problem which is described in chapter 9. Part of the problem involves the motion of a column of oil. The density of the oil is not provided in the data. Many of the student groups who have worked on this problem have asked for the value of this density. In fact the method of solution of the problem is sufficiently approximate that whether the specific gravity is taken as 0.8 or 0.9 makes little difference to the conclusion. The author has usually attempted to persuade students to overcome their lack of data by estimating a sensible value of specific gravity and then investigating how sensitive the result obtained is to small variations of the assumed value.

Such a process is of general applicability. If a physical value is needed and can be estimated, then use of the estimated value coupled with a sensitivity analysis will usually reveal whether the effort of obtaining a more accurate value is justified. If a 10% variation of the estimated value only alters the final outcome by say 2%, and 2% accuracy is acceptable for the purpose in hand, there is little point in obtaining a 1% accurate value of the value estimated. On the other hand, if a 2% variation of the estimated value changes the final outcome by 10% and this accuracy is not sufficient for the purpose in hand, the expenditure of the additional effort will be justified.

In the same way worst case analysis is often of great value. In many models some decisions about the type of model to be used can be taken, or some approximation can be made in the analysis, which can be shown to produce a lower or upper bound on the range of possible outcomes. If an approximate model shows that, in the worst case, the value of some resultant is within the range of acceptable values or that, in the best case, the resultant value lies outside the acceptable range, there is little point in investing great effort in constructing a more accurate model of whatever subject is being investigated.

3.6 Use of mathematical tools

The final general point to be made concerns mathematical aids and tools. In other contexts in tertiary education it is often educationally desirable to ask students to carry out computations using methods which may be inefficient or sub-optimal in some way. Usually this is for the purpose of enhancing understanding of some general principle or point in the mathematics being studied. In a modelling context, and particularly within the context of the case study course which is described in the next chapter, students should be encouraged to use all the mathematical aids available. This will encourage them to develop a familiarity with tools of the type which they will need to use routinely in the world beyond the university.

Many students, in the author's experience, only become familiar with standard mathematical reference works (such as Abramowitz and Stegun, 1965, and Gradshteyn and Ryzhik, 1965) through problems thrown up by modelling exercises. Modelling exercises also often produce a mathematical problem requiring numerical solution. This requirement, if appropriately dealt with, can be used to engender a familiarity with such tools as the NAG and GINO libraries and other standard computer subroutine libraries. Other computer packages, such as symbolic manipulation packages ('computer algebra'), dynamic system simulation packages, discrete event simulation packages, spreadsheets and statistical analysis packages may all be used by students in the course of a mathematical modelling course. Students will, of course, need either a pre-existing familiarity with these software tools or encouragement and assistance in locating and learning to use the appropriate items. Clements (1986a, 1986b) describes the uses to which a dynamic simulation system, BCSSP, described elsewhere by Clements (1984b, 1985), has been put in a mathematical modelling context. Moscardini, Cross and Prior (1984) also comment on the important role which simulation has to play in the practice of mathematical modelling.

3.7 Conclusion

This chapter has attempted to summarise a number of general points about mathematical modelling. Modelling is an activity in which experience can play as important a role as fundamental mathematical knowledge. All practitioners build up their own fund of experience and their preferred modes of operation. The points made in this chapter arise from the author's own experience and his observations of student groups tackling modelling problems. The conclusions drawn therefore represent, to a considerable extent, personal preferences and insights. Taken in this spirit it is hoped that they will have something relevant to say to other mathematical modellers, both experienced and inexperienced, about the art and science of mathematical modelling.

4

A course of case studies

The last two chapters have explored something of the background to, and the practicalities of, mathematical modelling. This book, however, is concerned primarily with one particular technique which has been used to develop, in students, the skills of mathematical modelling. In this chapter the development of the case study/simulation method is traced and its use by the author in his own teaching activities is described.

4.1 The context of the case study course

At the time when the course was designed, created and first used the Faculty of Engineering at Bristol University comprised five departments, Aeronautical Engineering, Civil Engineering, Mechanical Engineering, Electrical and Electronic Engineering and Engineering Mathematics. Historically the role of the Engineering Mathematics Department had been to meet the mathematics teaching requirements of the four engineering departments. Such an arrangement has considerable advantages over the possible alternatives whereby the mathematics teaching needs of engineering students are met either by service mathematics courses given by the Mathematics Department of the institution concerned or by the engineering departments themselves. In the first case there is often a strong feeling that mathematics lecturers have insufficient appreciation of and sympathy for the special needs of engineers resulting in rather theoretical mathematics courses which do not engage the interest of engineering students who, after all, are primarily interested in the applicability of the mathematics to their own disciplines. In the second case there is a tendency for the engineering lecturers to teach mathematics as a series of recipes for solving particular engineering problems with insufficient attention to the overall structure of mathematics as a subject in its own right. The solution adopted at Bristol had always been found to be a satisfactory compromise, the engineering mathematics lecturers being able to teach mathematics as a suitably coherent subject in its own right whilst giving sufficient attention to the applications of the subject to the engineering disciplines to engage the attention and interest of engineering students.

There had been, in the 1960s and early 1970s, an increasing realisation that, if mathematics was to fulfil its potential role in the engineering sphere, there was a need for mathematics graduates to have a better appreciation of the areas of application of mathematics in engineering and in industry and commerce. The staff of the Engineering Mathematics Department perceived that the Department was ideally placed to mount a degree course oriented towards fulfilling this need. As a result the design of a degree course in Engineering Mathematics was undertaken, the course was sanctioned by the University Senate and the first undergraduates commenced their studies in October 1977.

The degree program which resulted took as its objective the production of graduates with roughly the same level of mathematical skill as their counterparts on Applied Mathematics courses but with a background in and an orientation towards engineering rather than the more usual mathematical physics background commonly acquired whilst studying applied mathematics. To this end students on the course, during their first year, were to take a fairly wide range of general engineering courses. These were chosen from amongst those given by each of the four Engineering Departments to the students of other branches of engineering. At the same time the Engineering Mathematics students studied largely the same mathematics as students taking one of the four engineering degrees. Over the years this has been modified somewhat to make the first year of the course a little more challenging mathematically for the Engineering Mathematics students than it is for the other courses. During the second and third years of the course the mathematical pace was to be increased considerably, building on the sound engineering background provided by the first year.

The target group for recruitment of students to the course was those who might otherwise have read Applied Mathematics rather than those who would otherwise have read a more specific engineering discipline. It was hoped that the course would, in this way, increase both the quantity and quality of entrants to the engineering profession. In the event this objective has not, over the years, been fully realised. Many students on the course say that they chose it because they felt they had wanted to study engineering of some sort but were not yet certain which variety and so chose the course in order to keep wider options open. This, at first sight, does not seem to be widening the catchment of the engineering profession in the intended way. However it is probable that many of these students wishing to keep wider options open would, in the absence of the Engineering Mathematics course, have gravitated towards mathematics or even physics. Some, of course, would have found places on Engineering Science or General Engineering degree courses, but such courses are fewer and often less accessible than the more specialised engineering courses.

It should also be noted here that the concept of Engineering Mathematics envisaged for the course was a fairly wide one. It was intended to include all those branches of mathematics likely to find common applications in industry and commerce. For instance operations research topics, as well as being used in the management of a wide range of businesses including engineering ones, have a considerable overlap with the techniques of process control and control engineering and are included within the scope of Engineering Mathematics for our purposes. In the same way a large area of practical statistics (as opposed to the more abstruse areas of applied probability which are, in mathematics departments, sometimes referred to as statistics) is relevant to such topics as the quality control of manufacturing processes, the analysis of development trials data and the study of reliability and maintainability engineering. For this reason considerable statistics is included within the remit of Engineering Mathematics. In this way the mathematical net of the degree program is cast fairly widely and the variety of mathematics studied by its students is large.

4.2 The need to teach mathematical modelling

As we have discussed in chapter two, there was, during the 1960s and 1970s, an increasing realisation that traditionally trained mathematics graduates were not entirely meeting the needs of industry. (To make that statement might seem, at first sight, to be to accept that meeting the needs of industry is a prime function of university mathematics courses. Many within, and some without, the universities would not accept such a proposition. To turn the point on its head however, we might note that graduates who are well equipped to cope with the demands which their employments make upon them are the more likely to find their work fulfilling.) The work of McLone (1973), to which we have previously referred, brought out this point very strongly. The problem seemed to lie, at least partly, in one aspect of the structure of the traditional mathematics degree course – the examination system. Examinations and other assessments have a very strong effect on what students learn during a course. A prime student concern is, naturally enough, passing the assessment and, in most cases, passing it as well as possible. Students are, by and large, very efficient at devising strategies for optimising their performance in assessments and these strategies determine, to a very large extent, what they choose to learn and what they choose to ignore in a course. It is true to say, of most normal students, that no amount of exhortation by a lecturer will induce them to spend a great deal of time on learning skills or facts which they perceive to be of little utility for the purpose of the coming examination. This influence of the assessment procedures on the learning activities of the students is known as a 'backwash' effect of the examination or assessment procedure.

The connection between the curriculum and assessment methods which we have just noted was also explored by McLone (1971). His paper notes that traditional timed examinations were only able to assess a restricted subset of the basic mathematical abilities which, in his opinion, mathematics degree courses ought to be developing in students. He also asserted that a wider range of teaching methods was needed if all the desirable skills were to be taught.

The traditional university degree examination, in mathematics at least, consists of a number of examination papers, usually of three hours duration, containing a moderate number of self-contained questions. Each question is well posed, containing all the data needed for the solution of the problem, leading towards a single, well defined answer and taking, for the average student, usually no more than 45 minutes. The students have come to expect that a form of reciprocal principle is applicable to such questions – all the data needed is given and all the data given is needed. Knowing that such is the form of the examination, and having a reasonably well developed sense of responsibility towards their students, most lecturers provide, for each of their lecture courses, sets of examples or problems having much the same style and properties. Occasionally some lecturers will try to be more adventurous but, as those who have tried will recognise, such are the pressures of the assessment system that there is much resistance from the students to any attempt to try more adventurous and open-ended exercises.

The format and structure of the traditional university (and more latterly polytechnic) mathematics degree course then encourages students to perceive mathematics as being concerned not with the solution of the messy, ill structured, open ended problems which the real world provides, but with the solution of neat, well formulated, relatively restricted problems which have a single 'correct' solution method leading, after 30 to 45 minutes effort, to a well defined 'right' answer. This restricted view of the nature of mathematics has a deleterious effect on their activities as mathematicians once they leave university.

For engineering students this effect is extended to include a strong sense of compartmentalisation. Many engineering students view mathematics not as a subject which will illuminate their understanding of their engineering studies but as an additional obstacle placed in their path towards an engineering degree. The author has, from time to time, set problems on mathematics exercise sheets used in his mathematics courses for engineers which require, for their solution, the use of knowledge drawn from engineering courses as well as from the mathematics course. Such problems always raise considerable difficulties and, when the suggestion is made that some elementary application of engineering principles will aid the solution of the mathematical problems, many students express a sense of outrage (often expressed in the form 'But that's not mathematics' or even 'That's not fair!').

In contrast to the nature of undergraduate mathematics are the demands made on mathematicians working in industry. For them problems are rarely well defined. Typically problems arise and are presented to such mathematicians not in the language and terminology of mathematics, but in those of the sphere in which the problem arose, chemical engineering or management for instance. The task of these mathematicians is firstly to create a mathematical structure which adequately reflects or models the relationships of the entities in the source domain of the problem, or alternatively, perhaps, to perceive the underlying mathematical structure of the problem and make it explicit. Having done this they must then mobilise all their mathematical skills to arrive at a mathematical solution of the problem. The mathematical solution, however, is not the ultimate objective. That solution must then be expressed in the terminology of the original problem domain and, even more importantly, its implications for the original problem identified and described. Finally, the whole mathematical work must be explained to the original 'owners' of the problem in terms which they find understandable and intuitively acceptable.

In addition to this, industrial mathematicians will be facing, and working within, very different constraints from their academic counterparts. Non-mathematical considerations such as cost of solution, the computer and software resources available, access to library facilities and the urgency of completing the work all have an influence on the mathematical process of solution and must affect the mathematical nature of the model created. This is a very different situation from that which faced them as undergraduates. The constraints of the academic world and the academic criteria for an acceptable solution – elegance, economy of expression, compactness, preference for the analytic over the numerical – exert very different pressures on the solver. Typically academic training will have conditioned graduates to strive for the best possible solution (where best is measured in terms of the academic criteria mentioned) whilst the requirement on mathematicians working in industry is that they find not *the* answer but *an* answer which is adequate and acceptable *within the constraints pertaining*. Often the industrial mathematician will be led thereby to adopt solution techniques which the academic mathematician would shun as unsatisfactory or inadequate.

Another strong contrast is that the industrial mathematician is usually faced with a much more open ended problem than the undergraduate. Typically there will be a wide spectrum of possible models of the problem to be solved from which must be chosen one which will give an adequate answer for the purposes of the real problem. Choosing an efficient method which will yield an appropriate solution will give the modeller personal satisfaction but, provided the constraints of the organisation are met, this is not essential to the institution or company.

Cornfield (1977) gives an illustration of the different nature of the criteria by which the work of industrial mathematicians is judged. In describing the work of the Electricity Council Research Centre he gives four criteria which must be satisfied before any proposed innovation is acceptable in that industrial environment. Firstly, it must save money. An innovation which improves the quality of something which is already of adequate quality, no matter how little the improvement costs, is not useful. Secondly, it must fit in with present systems and practices, for most industries have an enormous investment in current plant, hardware, software and human resources and to throw all those away for a radical new system would rarely be justifiable. Thirdly, it must be simple to use by the practising engineers who will have to implement and operate it. Fourthly, it must be based on realistic comprehensive models of physical reality and have intuitively acceptable conclusions.

Another difference is that industrial mathematicians will typically be required to work not as individuals but as members of a team dedicated to the solution of the problem in the most advantageous way to the company or institution concerned. The working environment of the undergraduate is much more individually oriented, with students making their own lecture notes, working at their own problems in their own time and at their own pace, and ultimately standing or falling by their own efforts in the final examination.

In all these ways the demands made on industrial mathematicians are very different from those which the academic mathematician faces and for which their undergraduate training has prepared them. The quotations from Klamkin (1971), McLone (1973) and Gaskell and Klamkin (1974) given in chapter two all serve to reinforce the points which we are making here.

4.3 The objectives of teaching modelling

Having considered the problems of the mismatch between the skills engendered by a traditional mathematics degree course and those required by the working mathematician and decided that an attempt would be made to at least begin to teach the required working skills, it was necessary to choose a mode of teaching to achieve the objective. Firstly, though, some thought was given to defining more closely what was to be taught. The following set of objectives was formulated.

1. To give students practice in synthesising mathematical models from engineers' and other non-mathematicians' descriptions of physical and industrial systems.
2. To give students practice in interpreting the results of mathematical models in physical terms and critically evaluating their implications.
3. To give students practice in evaluating the effects of the various sections' of, and inputs to, models, and making appropriate simplifications and approximations to aid efficient solution.
4. To alert students to the non-mathematical constraints under which mathematicians operate in the industrial environment.
5. To give students practice in critically examining the various possible approaches to, and models of, a system, and choosing an optimal or near-optimal method of analysis within the constraints of the system.
6. To develop students' confidence in themselves as model builders and problem solvers.

These objectives could, like any other set of objectives, no doubt be criticised and improved. They were, however, taken as a starting point for the development of the case study course and have been effective in focussing attention on the aims of this teaching activity both during initial development and during periodic reviews subsequently.

4.4 Simulations, case studies and the hybrid concept

Having identified the nature of the differences between what a mathematics degree course teaches and what the effective mathematician in employment needs, we can now turn to the problem of how to develop in our undergraduate students the necessary skills. The educational technique which eventually evolved was a form of hybrid between the simulation and case study techniques.

Simulation as a teaching and learning tool is used in a wide range of professions and trades. Basically we might characterise it as the use of an imitation of reality to facilitate the training and improvement of performance of individuals when they face the real situation. The most commonly and widely known example of this is probably the flight simulator used to train military and civilian pilots. Similar examples are the simulators for the control rooms of such places as chemical plants, nuclear power generating stations and warships which are used to train, in the procedures and skills which they need, the teams of personnel who operate the real items. Military manoeuvres are another form of simulation exercise. There are, of course, much more lowly examples of the utility of the simulation method. In medical education trainee doctors practise the taking of medical histories and the making of diagnoses with simulated patients; shop assistants, DHSS officials and similar people are often trained in dealing tactfully with awkward or belligerent customers through the use of simulated encounters and schoolchildren may learn about world trade and development through playing the roles of the various nations in educational games simulating the international markets. A feature of the more sophisticated simulations is that they are adaptive, that is the situation simulated responds to the reactions of the trainees. Aircraft and control room simulators obviously must have this property.

Simulations are useful for a variety of reasons. Firstly they may, as in the case of the nuclear reactor or warship control room simulations, be used to train operators and commanders in procedures which will only be used in the sort of emergencies which it is fervently hoped will never arise. It is not possible to train operators for such eventualities using the real equipment. Aircraft simulators are sometimes used to train pilots in emergency procedures for situations which are, with appropriate training, survivable but in which the risks associated with the training are sufficiently great as to make training in the air on the real aircraft unacceptable. Simulators are often much cheaper to operate than the real equipment and so operators can economically be trained to much higher skill levels than might otherwise be feasible. Simulation can sometimes speed up real time or eliminate waiting or transit time and so allow trainees to achieve higher skill levels within a restricted time available for training. Simulation can be controlled and monitored in ways which are often unachievable on the real equipment or in the real situation. Trainers can therefore control the sequence of demands which are made on the trainees and monitor, measure and record their responses much more precisely and completely for subsequent analysis.

Learning in simulations takes two main forms. Firstly there is adaptive learning which takes place during the operation of the simulation and secondly there is the post-mortem learning which takes place when the trainers and trainees together review the progress of the simulation, possibly with the help of data recorded during the simulation. Different simulations rely on the two forms of learning to different extents but, in most simulations, there are elements of both. The adaptive learning taking place during the progress of the simulation is basically the accumulation of a fund of vicarious experience of operating the equipment or situation simulated. Such learning is valuable but is greatly enhanced by subsequent post-mortem analysis and discussion and the identification of the functional and dysfunctional elements of the performance and the ways in which the performance might be improved on future occasions.

In the previous sections of this chapter we have described the perceived mismatch between the skills acquired by undergraduate mathematicians and those needed subsequently for the effective practice of their vocation in the real world. It could not be denied, however, that many mathematicians in industry were, nonetheless, doing very effective work. How did this come about? Obviously well trained and intelligent mathematics graduates were proving sufficiently adaptable to learn the required skills 'on the job'. The experience of using their mathematical knowledge and skills in the solution of real problems was encouraging in them the development of the skills which we have identified above. This fact suggested that one possibly profitable approach to the teaching of such skills to undergraduate students would be to use simulation to give them false or vicarious experience which would, nevertheless, be sufficiently realistic to enable them to develop the skills of subsequently applying their mathematical knowledge to the real world. Such simulations would involve realistic problems presented to the students through a collection of the kind of documents which they might meet in industry – memoranda, correspondence, design drawings, data, technical notes and reports. The problems would be presented in the terms and terminology of those who might face such a problem and would include the context and some indication of the constraints involved. Students would be required to tackle the problems, devise solutions and report their results as they thought appropriate within the simulated environment.

The role of the staff in such teaching exercises would be to play the role of the head of department or section or team leader in the simulated industrial environment, that is to act as advisor, encourager, supervisor, manager and report receiving authority. In acting out such a role staff would obviously have to give considerable attention to acting within the simulated role and not merely continuing the normal tutorial function of university academic staff. The role needed to be one of drawing out from the students their ideas, encouraging their mathematical creativity, assessing, quickly, their proposed modes of tackling the problems, pointing out any immediately obvious problems or errors at a strategic level and generally acting as a source of experience and guidance but not offering detailed instructions nor minute checking of the mathematical work. This is a rather different role from that normally fulfilled by staff in tertiary education institutions and would obviously make demands which might be difficult to meet.

Where were these simulation exercises to come from? Obviously one possible source was from the imaginations of the author and his colleagues. Another more satisfactory source would be from industry itself. If industrial companies could be persuaded to donate examples of problems in the solution of which mathematics had had some significant role, these problems could be used as the basis for simulation exercises of the type outlined above. This approach would have the advantage that, during the post mortem stage of the simulation when staff and students put aside the roles they had been playing and reviewed together the progress of the simulation, the approach to the problem adopted by those originally facing it in industry would be known and available for comparison.

This possibility introduced into the course the second element of the hybrid concept, that of case study. Case study is the term usually used to describe the detailed study of the application of a particular theoretical idea or group of ideas to an actual practical situation. It should be differentiated from merely describing the application of theory to a generalised set of practical problems in that any specific situation will usually give rise to particular difficulties which have to be overcome if the application is to succeed (for instance the non-availability of all the data required to calibrate the theoretical model or deviations of components of the system from the properties of the idealised components which are assumed by the theory). Such difficulties will introduce the necessity for stratagems and minor modifications of the theory or its application. Case studies illustrate the necessity of making approximations and using judgement in the application of theory to the real world. They do not conceal the possible shortcomings of the particular cases studied. This description of the details of the particular application, warts and all, is what gives a case study its distinctive quality. It is obvious how this concept applies to the combined simulation/case study concept which was being developed here.

Having chosen this combined simulation/case study method for the course (for convenience a shortened title was chosen which, for historical reasons, was 'case studies' even though this term, on its own, is a slightly misleading title), the development of the details of its implementation revealed some further advantages. The first of these related to co-operative and group working skills. As we have already noted, industrial practice usually involves working as members of a team rather than as individuals. Undergraduate degree courses (with honourable exceptions) are not usually particularly appropriate training grounds for co-operative working skills. That this is so is evidenced by the emphasis which many prospective employers place upon investigating the extra-mural interests and activities of applicants for employment. Particular types of activity, such as participation in the training activities of the reserve armed forces or in team sports, are deemed to give an indication that those involved may have interpersonal skills and qualities of leadership and organisation which employers value.

The simulation method to be used for this course could be equally well used with individual students or with groups. If small groups of students were set to work on each case study they could be instructed to act as a team and produce a single report on their work. This would also have the advantage that somewhat larger and more realistic tasks could be given without making excessive demands on the time of any individual, and also that, working in teams, students could offer each other help and encouragement. This latter was felt to be important, particularly in the early stages of the case study course, since it was anticipated that students would, at least initially, find the modes of working demanded by this course very unfamiliar, confusing and probably somewhat frightening.

The second additional advantage related to the scope of the material which would form the basis of the case studies. At the time when the course was first mooted there was considerable concern in educational and industrial circles that industry was having difficulty in recruiting sufficient high calibre graduates for its needs. No single explanation of this problem was generally agreed to be totally convincing, but there was some feeling that industrial employment was suffering from a bad press and that graduates perceived such employment as dull, routine and unexciting. Changes in the political management of the country have succeeded in reducing the attractions of alternative employments to the point where graduates in the late 1980s no longer see industry in quite the same unfavourable light as their counterparts in the mid 1970s did. At the time, however, it was seen that the case study course, by presenting a wide range of interesting problems which had all come from industry, could help to alert students to the variety and interest of the mathematical applications which undoubtedly did exist within industry and commerce and, as a result, increase the attractiveness of such employment to graduates.

These beneficial side-effects, therefore, allowed further desirable educational objectives to be met by the case study course. Two further objectives were therefore formulated.

7. To give students practice in co-operative (as opposed to individual) working, and improve their group working techniques, and
8. To demonstrate the range of problems which are modelled mathematically, and illustrate the scope of the industrial mathematician's task.

4.5 The creation of the case studies

Having decided that the case studies were to be based, if possible, on problems which had been faced and solved by real companies and institutions, it was necessary to acquire the material. Published sources were used to compile a list of roughly 200 UK companies and governmental and non-governmental institutions which met two conditions; they had a significant annual expenditure on research and they employed a considerable number of graduate level mathematicians and engineers. The companies were approached, through their Directors of Research or similar officers, for assistance. The objects of the course and the needs of the University were explained and help in the identification of suitable problems and the provision of source material was sought.

Around 40 of those institutions approached responded favourably and, in many cases, with considerable enthusiasm. In most of these cases a meeting was suggested and, over a period of nearly two years, the author visited most of the 40 companies, explaining the requirements of the course in more detail and discussing possible problems suggested by the companies. The range of both problem types suggested and materials supporting the problem which could be made available to the University was very wide. Obviously, for the type of exercise envisaged, some details of the way in which the problem had originally arisen and been presented to the solver were highly desirable. Some information about the stages in the progression of the resulting study and the interim and final reports was also needed. In some cases much of this material was available in written form and in others it could be collected by interviewing participants. In some cases, though, most of the background had not, for one reason or another, been recorded or remembered and in these cases inference or even speculation was sometimes used to reconstruct the background.

After the potential problems had been discussed and material had been collected to support the most promising ones, the actual case study source documents were written. It had been decided, as a matter of policy, that all the case studies should be written up under fictitious names. This was requested by the donor organisation in some cases and in others the use of any single name might have been misleading since elements of more than one problem were included in some of the case studies. Whilst the intention of collecting real source material was to maintain the greatest degree possible of realism in the case studies it was, nonetheless, sometimes necessary to modify the problems, perhaps to exclude particularly obscure or abstruse mathematical elements or to reduce the scale of the problem. In many cases the original statement of the problem assumed a certain amount of knowledge which would have been common to those working in the particular environment but which would be quite unknown to a typical undergraduate. Such background had to be written into the case studies through some artificial device (such as a simulated couple of pages taken from a sales brochure or a memorandum explaining an institutional context for a recently arrived member of the research staff). Whilst such devices were artificial in an absolute sense they could, with skill and experience, be made to fit in with the spirit of the exercise and were used in that way in appropriate cases.

We mentioned in the last section the unusual demands made by exercises such as these on teaching staff. To assist staff to fulfil their role each case study was accompanied by a set of teaching notes describing the approach to the problem adopted by the donor organisation. Such notes also included comments on potential alternatives which might be suggested. These notes have been updated and expanded over the years particularly with respect to alternative solution methods offered, and have formed the basis of much of what appears in Chapters 5 to 11 of this book.

4.6 Practical considerations for the course

It was originally planned that the case study course should appear in the final year of the whole degree course. This positioning was suggested by the necessity for the students to have obtained an appropriate body of mathematical knowledge to use in their work on the case studies and a satisfactory degree of familiarity with that mathematical knowledge to be able to use it in this context. It was the author's expectation that mere knowledge of a piece of mathematical theory or technique would not, in itself, suffice to enable that theory or technique to be used in the creative manner demanded by this type of working. Rather, students would need to have had time to internalise that knowledge or technique and have become sufficiently familiar and comfortable with it to be confident of their ability to use or apply it. Subsequently, discussions with others involved in the teaching of mathematical modelling has revealed that this is a common expectation and the experience of the author and others bears out that the minimum satisfactory gap between students meeting a piece of mathematics and their being able to use it in exercises such as these is about twelve months. Use of the case study course even in the final year of the degree course would then imply that no mathematics beyond, at best, midway through the second year could reliably be presumed.

In the event, restrictions imposed by other factors in the design of the degree course dictated that the case study course should appear late in the second year. Although this was judged, a priori, to be sub-optimal, it turned out to be a surprisingly successful placement. One reason for this is the widening of the learning experiences which courses of this kind provide when compared with conventional lecture courses. It is observed that a wider range of learning experiences helps to improve student motivation and retain student interest. In the Engineering Mathematics degree course several teaching and learning techniques differing from the standard lectured course were planned to provide this widening. Some individual courses were chosen to be taught by guided reading. (The guided reading technique and its use in this degree course have already been mentioned in Chapter 1.) One such course appeared in the first year of the degree and two in the second year. The final year of the course included an extended project of a semi-research type. These activities both widened the learning experiences and helped to improve interest and motivation. The inclusion of the case study course in the second year resulted in at least one new learning experience being introduced into each year of the course. Experience has shown that the enthusiasm and motivation engendered by the case study course has more than compensated for the relatively restricted mathematical background which this placement implied.

Another benefit of the placing of the case study course in the second year of the degree is that many of the non-mathematical skills which this course attempts to teach – report writing, organisation of work, mathematical creativity and experimentation – are very useful to the students in their work on their final year projects.

Having decided, for the reasons mentioned earlier in this chapter, that students were to work on the case studies in groups rather than as individuals, it was necessary to decide upon the size of the groups to be used. If the groups were too small then both the advantages and the problems of group working would not be sufficiently apparent. If the groups were too large there would be the probability that some students would, either through lack of inclination or lack of opportunity, make little contribution to the work of the groups. Initially a group size of four was chosen. In the light of several years of experience with the course this has been reduced to three. With this size of group there is little opportunity for any member to take a passive role and pressures within the group to share out the not inconsiderable workload involved ensure that everyone has both the opportunity and the necessity to contribute. At the same time there are sufficient members (just) for the group to need to consciously organise its work patterns and to ensure that each member can contribute something which is commensurate with the individual's skills, whether these be a facility with computers and programming, an ease with model building and mathematical manipulation or a talent for technical writing and reporting. A related issue is how the groups are made up. There are two main alternatives. Firstly, students could be organised into groups by a member of the academic staff. Such organisation could be done along strictly alphabetical or other non-academic lines or by consideration of academic and other talents and skills with the objective of producing groups with balanced capabilities and talents. The second alternative is that students should be left to form their own groups. At Bristol the second alternative has been chosen. Experience shows that students are very shrewd at assessing their colleagues' capabilities and talents and have, for the most part, spontaneously organised themselves into quite well balanced groups. This option is probably only really viable with the fairly small student numbers on the Engineering Mathematics degree course and the first option would probably have to be taken with larger degree groups.

The course is organised around a series of meetings of the individual groups with a tutor who plays the role of a section leader, departmental head or other responsible authority within the institutional context simulated. These meetings take the form of reviews of progress on the study, interchange of ideas and suggestions for further action and receiving of final reports. Such meetings have been scheduled to take place once a week and each meeting is nominally of 45 minutes duration. This format has been found to be a practical compromise between the various possibilities. Progress on the studies could probably be made more rapidly with more frequent meetings but there is obviously a maximum frequency, probably twice weekly, beyond which meetings take place so often that the student groups have insufficient opportunity, in between formal meetings, to meet as a group without the tutor present to organise progress of the study. Students are told, at the beginning of the course, that, as in any industrial environment, their head of section (as played by the tutor) is a busy person but is always available, if he or she can be found, for brief consultations and advice between scheduled meetings. It is then up to students to use this resource to the best advantage they can. Experience again shows that groups who avail themselves of this offer, probably at some inconvenience to themselves, gain much in terms of the rapidity of their progress.

The case studies themselves were designed to be completed in about two or three weeks work each. This enabled each group to tackle a series of three to five case studies during a term and made it possible for each group to work on a representative range of different problems during the course. At the end of each study groups are required to produce a report. In keeping with the simulated industrial environment these reports are varied. In some cases a full technical report is required, in others a short note only. Other report formats which have been requested include a verbal presentation on the work to be made to a 'client' who has commissioned the study and a single side briefing note for the benefit of the head of section who has to report on the work at a senior management meeting. The variety of report formats used helps students to develop a critical approach to report writing, tailoring what they produce to their perception of the needs of the audience addressed.

At the start of the course a single lecture is given in which the philosophy of the course is introduced, some points are made about the day to day organisation of the course, some introductory ideas about the practicalities of mathematical modelling are discussed and some of the mathematically peripheral activities which the studies will demand are highlighted. Some consideration has been given, over the years, to one or more lectures on the subject of mathematical modelling being given at the end of the course. Such lectures would enable some loose ends to be tied together and some lessons to be drawn from the experiences which the students have, by that point, accumulated. Such lectures have not, to date, been implemented but may be introduced in the future.

4.7 Student reactions to the course

It is quite clear that the demands made upon the students by this type of course are very different from those made by more conventional lecture courses. How do the students react? The course structure described above dictates that, after a single initial lecture introducing the course, the students are rather thrown in at the deep end. As a result it is to be expected that they will initially be rather confused about what exactly is expected of them and probably, to some extent, frightened by it.

This expectation has indeed been borne out in practice. The typical pattern of progress is that after the first week working on the first case study a very dejected group of students appear for their meeting with the tutor. It is usually the case that these groups are markedly reluctant to actually commit themselves to doing anything but, instead, are seeking the security of being told by the tutor exactly what to do before starting. The undefined nature of the problem and the lack of the usual cues as to which piece of mathematics should be used for its solution contribute to a sense of confusion and disorientation. It is necessary for the tutor, at this point, to try to draw out from the students some initial ideas about the ways in which they might progress and encourage them to follow up such ideas. Some ideas have usually already occurred to members of the group but, it is observed, there is considerable resistance to mathematical experimentation, students are reluctant to try out possible lines of mathematical progress before they are clear where such experimentation might lead. Hence such ideas as might have suggested themselves have usually been left unexplored. Considerable encouragement is needed to entice students to make tentative steps towards possible solutions before they feel they have a complete route to a solution mapped out. With encouragement most groups can be persuaded to take one or more of their own ideas and see where they lead.

The confusion and fear then rapidly disappear when the students begin to find that they are making progress. One of the most satisfying aspects of teaching this course is observing the way in which the confidence which students feel in themselves as model builders and creative users of mathematics grows with even their first minor successes. As groups embark on their second and third case studies they usually show much less reluctance to mathematical speculation and a greater preparedness to feel their way forward a step at a time. In fact, in some cases there rapidly develops a mathematical over-confidence and a somewhat over casual tendency to assume that any idea must be a good one. At this point it is necessary for the tutor to become perhaps less encouraging and more critical and to encourage students to develop in themselves a more evaluative and critical approach to their own creativity. Successful model building requires not only mathematical creativity but also mathematical judgement and the ability to choose sensibly between alternative possible avenues of exploration.

One of the lasting benefits of having a course such as this in the degree structure is the affective changes brought about in the students, that is the changes in their attitudes to mathematics and its uses and their perceptions of their own abilities with mathematics. It is gratifying to be able to report that the case study course is frequently mentioned by graduating students as one of the high points of the course and often one which has significantly changed their attitudes to the use of mathematics. It is difficult to quantify such benefits but they are by no means negligible.

4.8 The demands made on tutors

The skills needed by staff acting as tutors on a course such as this appear, at first sight, to be very different from those needed for lecturing. Staff need to be able to act within a role which requires that they react to student suggestions as though they were meeting the problem for the first time. They must largely put out of mind any preconceptions they have about the appropriate way to solve the problem in hand and the knowledge which they will inevitably have about the way in which those originally facing the problem reacted. They must relinquish their traditional role of source of knowledge and adopt, instead, the role of fellow seeker after a usable solution to the problem facing students and supervisor together. At the same time the role is that of a more senior and experienced member of the organisation, so it is perfectly in character for tutors to be able to offer advice from a position of greater knowledge, experience and maturity. What must be avoided is dictating to the students where they should go or by what means they should proceed. To do that would be to destroy much of the value of exercises such as this. To fulfil the role described requires an abandoning of the didactic role and a willingness to embark upon each meeting with no knowledge of where the quest will lead. This requires considerable self confidence and a preparedness to place oneself in a rather more exposed position in the presence of students than most lecturers are used to.

The role playing requirement is not as far from the skills of a good lecturer as might, at first sight, appear. To play a role satisfactorily requires an ability to think oneself into someone else's position and react as one supposes they would. A good lecturer and teacher already possesses such an ability in a different context. To teach effectively requires that the teacher is able to understand the problems of the student. The material which a lecturer teaches is, for the most part, familiar to the lecturer. To present the material in a manner which is understandable to students requires that the teacher is able to place himself or herself in the position of the student and anticipate which parts of the material will give the student problems of understanding and, further, to devise ways of explaining the material which will make its meaning and import clearest to the students at their state of knowledge and familiarity with the material. Further, a good lecturer will possess and use the skills of public oratory which are also part of the stock-in-trade of the acting profession. Such abilities are therefore not far from the skills of the competent role-player.

In general then experience suggests that the skills needed by a tutor on a course such as this can be developed by many lecturers provided they have first analysed the differences in teaching requirement which the course presents and keep those differences in mind whilst tutoring the course. It should perhaps be recognised that there are probably some for whom such skills might be particularly alien and difficult to acquire and such lecturers might perhaps be excused for preferring to remain in more conventional teaching roles.

4.9 Assessment of student performance

Students' ability and knowledge at the end of most university courses is assessed by some means. The most common method of assessment is the three hour unseen examination. The undesirable backwash effects of examinations have already been mentioned. In the case of an unconventional course such as the case study course consideration must be given to how, if at all, student achievement on the course can be measured.

The reasons for assessment are twofold. Firstly, there is an external demand that universities should provide some judgement of the relative abilities and achievements of their graduates. Such judgements are commonly used by potential employers as one of perhaps a number of criteria on which to base their choice of employees and their expectations of how recruits might be expected to perform subsequently. Indeed universities themselves use such judgements to choose those students who are qualified for further study for higher degrees. Secondly, there is the related issue of student motivation. Whilst, in an ideal world, we might wish that students pursued their studies purely for the love of the learning they acquired, it is undeniable that the university environment offers many inducements to students to neglect their studies to a greater or lesser extent. The knowledge that, at the end of the term or year, there is an examination which must be passed before they can proceed to further study helps to concentrate their energies and provides a control mechanism which ensures that an appropriate balance between academic studies and extra-curricular activities is, in most cases, achieved. Of course the examination system can also distort the academic studies of the students. The request which students commonly make to have clearly defined which parts of a course are examinable and which are not is usually a precursor to their neglecting the non-examinable (and often more interesting and significant) parts of the course.

One of the decisions which was taken early in the design of the case study course was that considerations of assessment would not be allowed to distort the design of the course. Since the objectives of the course were both cognitive and affective, and the cognitive objectives were fairly complex ones more in the realms of synthesising and evaluative activities, designing a conventional examination to assess the achievement of such objectives would be extremely difficult if not impossible. The option of a written examination was therefore rejected. The motivational aspects of assessment were met fairly easily by making satisfactory completion of the course a requirement for passing the year. Obviously requiring no more than a satisfactory completion of the course invites students to attempt to put in the minimum amount of work and concentrate their efforts on other, superficially more rewarding (in terms of examination credit) aspects of the course. In practice the satisfaction and enjoyment which students have found in the course once the initial confusion and bewilderment have been overcome have ensured that considerable effort has been evident in almost all cases.

Nonetheless an effective assessment scheme was judged desirable if for no other reason than the desire to be equitable in rewarding student effort. Over the years a suitable scheme has been devised based upon continuous assessment using the subjective judgement of tutors. Tutors are asked to keep brief records of the progress of each group week by week. Such records need only be very brief, perhaps a sentence or two, but are accompanied by a grade for general effort and achievement. Where a report of some form, written or verbal, has been requested an assessment of that report is included in the grade awarded. Such a grade could be kept on any reasonable scale, a mark out of ten or a grade on a scale from $\alpha +$ to $\gamma -$ for instance. At the same time a note is made of any particularly worthy contribution by any individual in the group and also any evidence that individuals have contributed little or nothing to the group's efforts. At the end of the course the group grades are averaged and the individual grades are used to award a slightly higher or lower grade to particular individuals in groups as deserved. It is noticeable that in groups which are well integrated and which operate well as groups there are rarely reasons for giving individuals marks which differ from the group average. Different marks are usually given to individuals who have struggled to excel in groups whose general performance has been rather poor. The marks from the case study course have then been integrated into the overall mark for the year along with other course work, continuous assessment and laboratory work marks.

The problems of assessing student achievement in courses of the mathematical modelling type have exercised a number of authors writing on modelling and modelling courses. Many of the papers in Berry *et al.* (1987), describing mathematical modelling courses which are in use in various tertiary education institutions, mentioned schemes which have been used for assessment. Oke (1980), describing a mathematical modelling course which was developed for an in-service MSc course in Mathematical Education, specified a mixture of self-assessment, coursework, examinations and project work which was used to assess that course. Usher and Simmonds (1987) described, amongst other topics, the assessment of undergraduate group modelling exercises. Their scheme involved both examinations and coursework. The coursework mark took into account written reports and oral presentations made by the group and a subjective mark awarded by tutors on the basis of their observation of group activity. They also made an unusual innovation in that some small weight was also given to an assessment made by individual group members of the contributions of their colleagues to the group effort. It is evident that the assessment of modelling and other activities similar to the case studies described in this chapter is a topic of some interest and much work remains to be done in this area.

4.10 Other similar developments

The work described in this chapter has occupied a considerable part of the author's working life. It has been described in part and in varying degrees of detail in various previous publications (Clements and Clements (1978), Clements (1978), Clements (1982a) and Clements (1984a)). That the idea is practicable and desirable is evidenced not only by the author's experience but by the adoption of the idea by others and also by the independent development of a number of similar initiatives. The most strikingly similar work of which the author is aware is that done by Agnew and Keener at Oklahoma State University, USA (described in Agnew and Keener (1980), Agnew and Keener (1981) and Agnew, Keener and Finney (1983)).

PART II

5

Andover Aerospace Components

5.1 Andover Aerospace Components – source documents

---

andover aerospace components ltd

to   T.Fellowes   **date** 12th May
          Simulation Group

from P.A.F.Doughty   **ref**   D/835/PAF/HY
          Head of Simulation Group

<u>Leda release gear</u>

This department has been asked to model the dynamics of a proposed
release gear for the Leda launcher.   I am attaching memos and a copy
of TN 714 which explain the proposal.   I have sent a memo to Arthur
Bolingbroke asking him for dimensions, weights etc. of the release gear
components and he should be sending these to you direct.   Could you
please investigate the dynamics of the release gear and, if possible,
determine the time taken to·retract the release gear with the proposed
jack.   If you need any more data please contact Arthur direct.

Memorandum 1

andover aerospace components ltd

to   P.A.F.Doughty   **date** 10th May
          Head of Simulation Group

from J.O.Gaunt   **ref**   D/835/JOG/LA
          Leda Project Manager

Jack sizing for Leda release gear

It has been proposed by Arthur Bolingbroke that we modify the old Orphus IV release gear to suit the new Leda Satellite launcher, thus effecting economies in both capital cost and construction time. He has outlined the necessary structural modifications in the attached Tech.Note. The component weights are, of course, changed by the structural modifications and the extra complication of the two link bracing arms has been added. We expect to need a larger capacity retraction jack for the gear and Arthur has proposed a Hadley compressed spring/hydraulic jack of 3250 lbs f capacity.

This jack size proposal is made on the basis of experience and judgement only and, since the retraction time for the Leda launcher release gear is relatively closely circumscribed (see Arthur's report) I would like some confirmation of its suitability if possible. Will your department look into the possibility of modelling the dynamics of the release gear retraction and providing confirmation of the suitability of the proposed jack?

Memorandum 2

andover aerospace components ltd

to   A.Bolingbroke   **date** 12th May
        Leda Project Team

from   P.A.F.Doughty   **ref**   D/835/PAF/HY
        Head of Simulation Group

Leda release gear

The Leda Project Manager has asked my department to attempt to
model the dynamics of the Leda release gear you have proposed in
TN 714 and confirm your jack sizing estimate.   I am putting Tim Fellowes
onto the task but he will need more information on the weights, dimensions
etc. of the components of the release gear.   Could you please send all
this information to him direct?

Memorandum 3

andover aerospace components ltd

to   T.Fellowes
           Simulation Group

date 14th May

from A.Bolingbroke
            Leda Project Team

ref   D/835/AB/LA

<u>Leda release gear</u>

     I attach a copy of figure 2 of TN 714 on which the relevant
dimensions of the proposed modified Orphus Launcher release gear
are shown as requested by Dr.Doughty.   If you need further
assistance or information please ring me on Int.593.

Memorandum 4, page 1

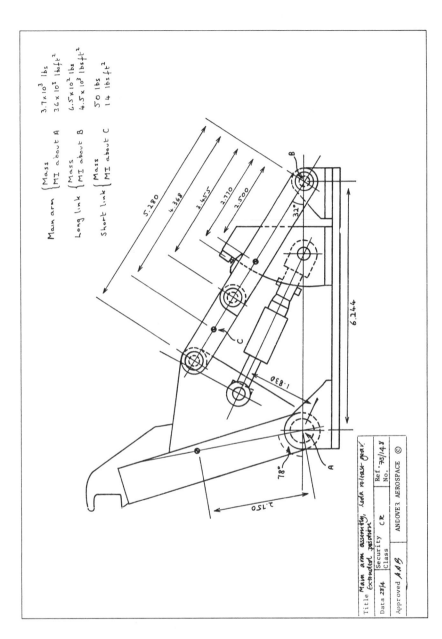

Main arm $\begin{cases} \text{Mass} & 3.7 \times 10^3 \text{ lbs} \\ \text{MI about A} & 36 \times 10^3 \text{ lbs-ft}^2 \end{cases}$

Long link $\begin{cases} \text{Mass} & 6.5 \times 10^2 \text{ lbs} \\ \text{MI about B} & 4.5 \times 10^3 \text{ lbs-ft}^2 \end{cases}$

Short link $\begin{cases} \text{Mass} & 50 \text{ lbs} \\ \text{MI about C} & 14 \text{ lbs-ft}^2 \end{cases}$

5.280

4.368

3.455

2.770

2.500

6.244

1.830

2.750

78°

32°

Title Main arm assembly, soda release gear, Extruded position

Data 28/4 | Security Class | CR | Ref. 75/148
| | No.

Approved AAB | ANDOVER AEROSPACE ©

Memorandum 4, page 2

andover aerospace
components ltd

TECHNICAL NOTE

title: PROPOSED MODIFICATIONS TO
THE ORPHUS IV LAUNCHER
RELEASE GEAR TO ENABLE ITS
USE FOR THE LEDA LAUNCHER

no: 714

issue: 1

date: 4 May

written by: A Bolingbroke
Leda Project Team

approved by: J O Gaunt
Leda Project
Manager

Technical report, page 1

1.   INTRODUCTION

    The Leda satellite launching rocket has been designed for positive
hold-down during the motor start-up phase. Four lugs are provided on the
launcher lower body and these will be engaged by claws on the release gear.
The release gear is maintained rigid during the start up of the launcher's
motors so that it is prevented from lifting off the pad until full thrust
has been attained. The release gear is then retracted and the launcher leaves
the pad.

2.   THE PROPOSED RELEASE GEAR

    The release gear used for the Orphus series of launchers is available
and could, with some modifications, be made suitable for the Leda system.
The Orphus gear consisted of four main arms whose function was to hold the
launcher upright before launching. The Leda release gear must perform this
function and also exert the positive restraint during motor start-up.
The main arms of the release gear (see figure 1)were considered floating,
being held against the launcher body by the hydraulic retraction jacks only.
The modified design adds a supporting link system which, when the arms are
in the extended position, are pushed slightly over centre upwards against
a stop. The hydraulic jacks are replaced by compressed spring, hydraulically
damped jacks whose tension, in the extended position,hold the links against
the stops and maintain the release gear rigid (see figure 2).

    To retract the gear the supporting links are pushed over centre downwards
and the combined effect of the compressed spring jack and the weight of the
supporting links then complete the retraction of the main arms (see figure 3).
The initial push is achieved by the operation of a standard Mk IV bomb release
gear which is fitted to each link stop.

3.   THE JACK CHARACTERISTICS

    The retraction time of the main arms will be governed by a number of
factors - the masses and moments of inertia of the moving components and the
forces exerted by the main jacks and the bomb release gears. The bomb release
gear exerts an approximately constant force over its stroke whilst the jack
exerts a linearly decreasing force over its stroke. The force/stroke
characteristics of the Mk. IV bomb release and a typical compressed spring jack
are shown in figure 4.

4.   THE PERFORMANCE REQUIREMENT

    The main arms of the release gear are required to reach the fully retracted
position shown in figure 3 no more than 0.65s after the operation of the bomb
release gear. It would be undesirable operationally for retraction to be too
rapid as this would tend to exacerbate the effects of any mis-sychronisation
between the four arms of the release gear. Additionally faster retraction calls
for larger and more expensive main jacks. It is estimated, on the basis of
experience in the operation of the original Orphus release gear, that a Hadley
spring/hydraulic jack of 3,250 lbs f capacity will be adequate to satisfy the
performance requirements. This jack has a force/stroke characteristic essentially
similar to that of figure 4(b) with a maximum force, F_0, of 3,250 lbs f
decreasing linearly to zero over an operating stroke, S_0, of 5·4 ins.

5.   RECOMMENDATION

    It is proposed that the Orphus launcher release gear is modified as
shown in figure 2 to suit the Leda launcher. The use of the existing
gear will effect economies of time, materials and cost. It is estimated
that a Hadley spring/hydraulic jack of 3250 lbs f. capacity will be
adequate to meet the performance requirements.

Technical report, page 2

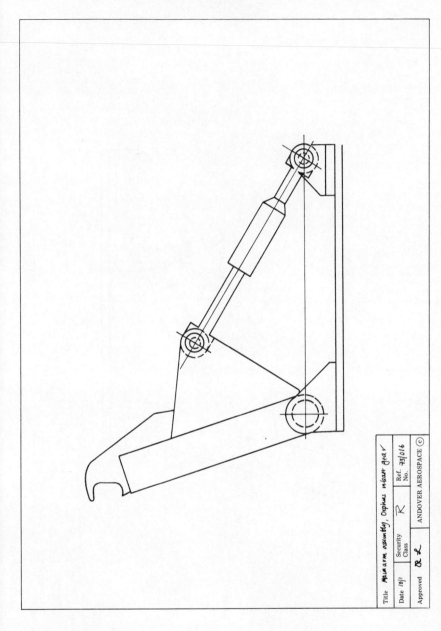

Technical report, page 3

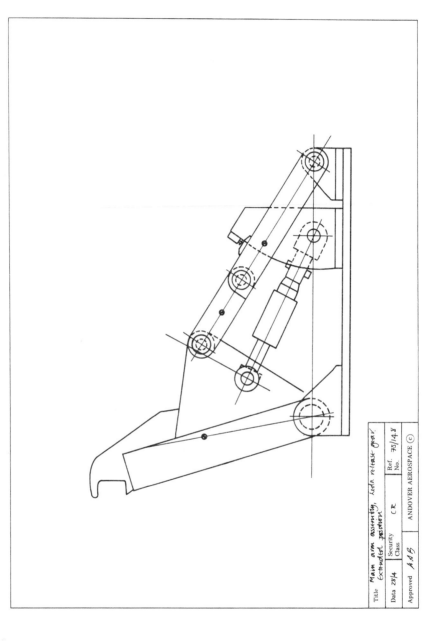

Technical report, page 4

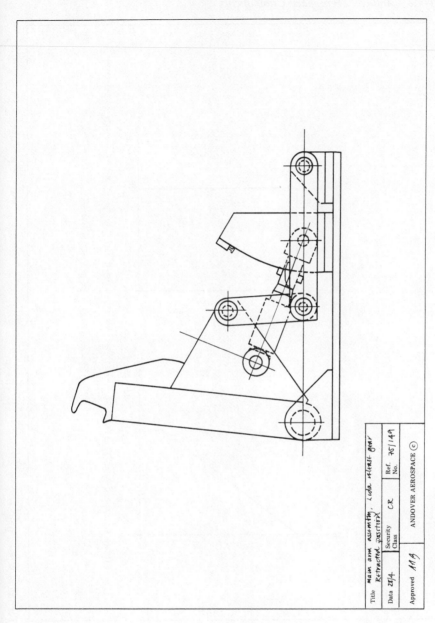

Title	Main arm assembly, load release gear Retracted position		
Data 25/4	Security Class	CR	Ref. No. 75/149
Approved MB		ANDOVER AEROSPACE ©	

Technical report, page 5

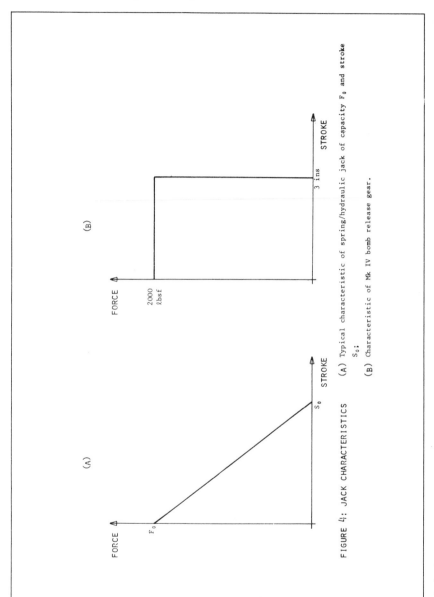

FIGURE 4: JACK CHARACTERISTICS

(A) Typical characteristic of spring/hydraulic jack of capacity F_0 and stroke S_0;

(B) Characteristic of Mk IV bomb release gear.

5.2 Andover Aerospace Components – general lines of donor's solution

The problem facing the Andover Aerospace Company's engineers is fairly fully explained in the documentation. Their response was to create a model of the release gear using fairly straightforward mechanics. The main decisions which had to be taken during this stage of the modelling related to what approximations might reasonably be made and which physical effects might be neglected. The modelling resulted in a differential equation governing the retraction dynamics of the release gear which was solved by a simple numerical method. The numerical result of this work then led, in turn, to a re-assessment of the adequacy of the retraction jack originally proposed and a recommendation for a larger jack.

The release gear was modelled as a three element system acted upon by external forces (the bomb release jack which initiated operation and the compressed spring jack which sustained it). The motion of the system was then obtained from the energy principle

> Initial potential energy + work done by external forces
> − frictional losses = current potential energy   (5.1)
> + current kinetic energy

Since the system was initially at rest the initial kinetic energy was zero and did not appear in the energy principle.

Two assumptions were then made to simplify the problem. Firstly the frictional losses were assumed negligible and secondly, noting that the mass and moment of inertia of the short link were considerably smaller than those of the other components, the rotational kinetic energy of that link was ignored and its translational kinetic energy and potential energy were approximated by placing an equivalent mass at the end point of the long link. Hence, in the notation defined by figure 5.1, the following hold

$$\text{current potential energy} = m_1 g h_1 \sin(\theta) + (m_2 h_2 + m_3 l_2) g \sin(\phi) \quad (5.2)$$

$$\text{initial potential energy} = m_1 g h_1 \sin(\theta_0) + (m_2 h_2 + m_3 l_2) g \sin(\phi_0) \quad (5.3)$$

$$\text{current kinetic energy} = \tfrac{1}{2} I_1 \dot{\theta}^2 + \tfrac{1}{2}(I_2 + m_3 l_2{}^2) \dot{\phi}^2, \quad (5.4)$$

where I_1 and I_2 are the moments of inertia of the main arm and the long link about their pivots.

The work done by the external forces was approximated by assuming that the forces acted at all times perpendicular to the radial lines from the pivot points at ground level to the point of application. With this assumption

$$\text{work done by external forces} = \int_{\theta_0}^{\theta} F_1 r_1 \, d\theta - \int_{\phi_0}^{\phi} F_2 r_2 \, d\phi.$$

It may be inferred from the graphs at the end of the technical note that F_2 is

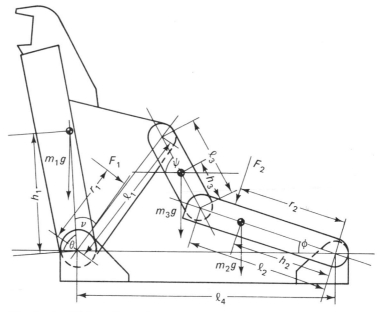

Fig. 5.1. Definition of notation.

constant until the end of its stroke, s_2 say, whereas F_1 is linearly decreasing. Thus

$$F_1 = F_1'[1 + (\theta_0 - \theta)r_1/s_1]H(s_1/r_1 + \theta_0 - \theta)$$
$$F_2 = F_2'H(\phi - \phi_0 + s_2/r_2),$$

where $H(x)$ denotes the Heaviside unit step function on the variable x. Hence

work done by external forces =
$$\begin{aligned}
&r_1F_1'[(1 + r_1\theta_0/s_1)(\theta - \theta_0) - (\theta^2 - \theta_0{}^2)r_1/2s_1]H(s_1/r_1 + \theta_0 - \theta) \\
&- r_2F_2'(\phi - \phi_0)H(\phi - \phi_0 + s_2/r_2) \\
&+ \tfrac{1}{2}s_1F_1'H(\theta - \theta_0 - s_1/r_1) + s_2F_2'H(\phi_0 - \phi - s_2/r_2).
\end{aligned}$$ (5.5)

Equations (5.1) to (5.5) combined to yield an ordinary differential equation of the form

$$f_1(\theta,\phi) = f_2(\theta,\phi) + \tfrac{1}{2}I_1\dot{\theta}^2 + \tfrac{1}{2}(I_2 + m_3l_2{}^2)\dot{\phi}^2.$$ (5.6)

This was not soluble without some constraint to link $\dot{\theta}$ and $\dot{\phi}$. The constraint was provided by the geometrical closure of the system. This gave

$$l_1\sin(\theta + v) = l_2\sin(\phi) + l_3\sin(\psi)$$
$$-l_1\cos(\theta + v) + l_2\cos(\phi) + l_3\cos(\psi) = l_4.$$

Eliminating ψ yielded

$$l_4{}^2 + l_1{}^2 + l_2{}^2 - l_3{}^2 = 2l_2l_4\cos(\phi) - 2l_1l_4\cos(\theta + v) + 2l_1l_2\cos(\theta + v - \phi).$$ (5.7)

Differentiating with respect to time resulted in a relation between $\dot{\theta}$ and $\dot{\phi}$,

$$l_1[l_4\sin(\theta+v)-l_2\sin(\theta+v-\phi)]\dot{\theta}=l_2[l_4\sin(\phi)-l_1\sin(\theta+v-\phi)]\dot{\phi} \quad (5.8)$$

that is

$$\dot{\theta}=f_3(\theta,\phi)\dot{\phi} \quad (5.9)$$

say. Substituting equation (5.9) into equation (5.6) gave

$$\dot{\phi}=-\left\{\frac{2[f_1(\theta,\phi)-f_2(\theta,\phi)]}{I_1f_3{}^2(\theta,\phi)+I_2+m_3l_2{}^2}\right\}^{\frac{1}{2}} \quad (5.10)$$

and

$$\dot{\theta}=f_3(\theta,\phi)\dot{\phi}. \quad (5.11)$$

Equations (5.10) and (5.11) could now be integrated by any suitable method for the numerical integration of first order ordinary differential equations (the Runge–Kutta method for instance).

The parameters needed, taken from the memo from Arthur Bolingbroke, were

$l_1 = 3.295$ ft	$l_2 = 3.455$ ft	$l_3 = 1.825$ ft	$l_4 = 6.244$ ft
$h_1 = 2.750$ ft	$h_2 = 2.500$ ft	$r_1 = 1.830$ ft	$r_2 = 2.770$ ft
$m_1 = 3700$ lbs	$m_2 = 650$ lbs	$m_3 = 50$ lbs	

$I_1 = 36\,000$ lbs ft^2	$I_2 = 4500$ lbs ft^2	
$F_1' = 3250$ lbs f	$F_2' = 2000$ lbs f	
$s_1 = 5.4$ ins	$s_2 = 3.0$ ins	
$\theta_0 = 78°$	$\phi_0 = 32°$	$v = 44°$.

Notice that Bolingbroke has actually forgotten to mark in the dimension l_1. This omission often causes students working on this problem to complain that without this dimension they cannot proceed. In the real world there would be two possible responses to this, either to telephone Bolingbroke and ask him for the missing dimension or to make a sensible estimate. The diagram is obviously a scale drawing so the measurement can easily be inferred from the drawing. If the tutor uses this little problem wisely students may learn a lesson about the real world – sometimes a lack of data can be overcome by the use of judgement and common sense!

The results of integrating equations 5.10 and 5.11 are shown in figure 5.2. It can be seen that a retraction time of around 0.67 s was indicated. This was not rapid enough to meet the criterion mentioned in the technical note. If the effects of friction in the bearings in the device had been accounted for the retraction would have been slower still so it was concluded that a stronger jack was needed to retract the device rapidly enough. In practice, of course, the sizes of jack which can be used are restricted to a range of standard sizes available from the suppliers. In this instance a series of further simulations would have been carried out to assess the various potential alternatives and a final choice made in the light of these predictions (and probably other considerations).

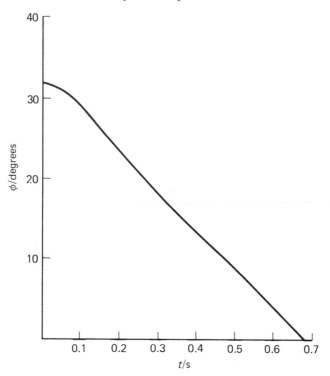

Fig. 5.2. Retraction history of release gear.

5.3 Andover Aerospace Components – commentary on student solutions

The first hurdle students have to surmount in this problem is the decision whether to approach it using an energy principle or using forces and moments. In the author's experience, those groups of students who have attempted the latter approach have invariably become submerged in a mass of equations and have lost direction as a result. One group once told the author that they had derived a set of more than forty linked linear equations describing the system by considering the mutual forces between the component parts and the equations of motion of each part!! There is little doubt that an approach via an energy principle, as adopted by the donor, is preferable.

It is worth pointing out here that the donor's solution made a fairly sweeping approximation at an early stage, the representation of the short link as a point mass located on the end of the long link. It is probably no coincidence that this approximation is exactly what is needed to eliminate any dependence on ψ from equation (5.6). It seems unlikely that this is the result of fortuitous luck or guesswork. More likely it is an example of the complex forward and backward interplay between stages of the modelling process described in chapter 2. The original solver of this problem probably reached a more complex form of equation (5.6), then backtracked to see what further approximations could be made to simplify equation (5.6) and ease its solution. It is also worth pointing out that formulating equation (5.9) with $\dot{\theta}$ as subject rather than with $\dot{\phi}$ as subject is probably no accident either. If formulated in the other way numerical instability in the solution of the differential equations may result near to the starting time since θ has a stationary value at this point. This is due, physically, to the short and long links being in line at the starting condition.

The next hurdle that has given almost all student groups trouble is the use of an implicit relation between θ and ϕ (equation (5.7)) to obtain a relation between $\dot{\theta}$ and $\dot{\phi}$. Many groups have obtained an equation equivalent to (5.6) (though usually containing ψ as well) but have been unable to proceed further because they cannot obtain explicit relations for two of the angles θ, ϕ and ψ in terms of the third. The use of an implicit relation in a computational approach appears to cause conceptual difficulties.

For those who, with or without prompting, get this far there is the choice between integrating the differential equations for both θ and ϕ (or possibly for θ, ϕ and ψ if a simplifying assumption has not been used to remove ψ from the problem) or integrating for one angle and obtaining one or both other angles by solving the implicit equation relating angles in what is essentially a four bar linkage, that is, for instance, obtaining θ not from integrating the differential equation but directly from equation (5.7). Since ϕ has been obtained from the differential equation, equation (5.7) can be solved for θ by Newton–Raphson or some equivalent equation solving procedure. Student groups have successfully adopted both of these approaches and there is little to choose between them.

Most student groups do not immediately decide to make the simplifying assumptions made by the donor which resulted in the elimination of ψ and $\dot{\psi}$ from the problem. As a result their equivalent equations to equations (5.2), (5.3), (5.4) and (5.6) are more complex, but the problem may still be solved as outlined. For those groups who, with or without prompting, decide to make such a simplifying assumption there is then the decision as to where to locate the lumped mass of the short link. There are more options than the one adopted by the donor. A sensible approach is to try out several possible simplifications and compare the results. The mass of the short link may be placed at the end of the long link, on the main arm, or distributed in some proportion between the two. In the event, trying out each of these options produces a change of no more than about 2% in the retraction time. Bearing in mind the gross simplifications that have already been made (neglect of friction, line of action of external forces et cetera) the result cannot be expected to be very precise so this level of uncertainty is perfectly acceptable. This is an example of the sensitivity check approach to approximations which was mentioned in chapter 3. In this case it is the nature of an approximation that is being varied to determine how crucial it is to the result rather than a parameter value but the principle is the same.

6

The Leather Working Machinery Group

6.1 The Leather Working Machinery Group – source documents

THE LEATHERWORKING MACHINERY GROUP

MEMORANDUM

To: Mr. T. Fellows   From: P. Franks
                                                 Group O.R. Department

Copies: Date: 16th July

Leasing Contracts and the Effects of Inflation

    I attach a copy of a Memorandum from Mr. Dixon, Head of the
Commercial Department of one of the Group's subsidiary companies.
Please would you investigate the problem raised and write a
report for Mr. Dixon within the next two weeks.

Memorandum 1

ENGLISH GLOVE MACHINERY COMPANY LIMITED

MEMORANDUM

To: Mr. P. Franks
     Group Operational Research Department
     Leatherworking Machinery Group

From: A.F. Dixon
       Commercial Department

Copies:

Date: 14th July

Leasing Contracts and the Effects of Inflation

I am concerned about the possible effects of the high inflation rates we have recently been experiencing on the rate of return on our machinery leasing business. As you probably know, our company produces the machinery which glove manufacturers use in their operations. Our business is traditionally founded on the leasing of these large and expensive machines to the manufacturers. Our standard terms of lease are as follows:

1. the annual rental of a new machine is set at $\frac{1}{8}$ of its new value, and that rental applies for 8 years;

2. at the end of the 8 years, the lease may be renewed for an indefinite period at a rental which is set at 7.5% of the rental for an equivalent new machine *at the time of renewal*.

As only minimal technical change occurs in our industry, and the machinery typically remains perfectly serviceable for up to 20 years or so, most customers accept a second lease period and only change to a new machine after 12 to 15 years. In times of small inflation the level rental system seems satisfactory, as it is popular with the customers and causes minimum administrative effort for us. Our rate of return also seems satisfactory at about $7\frac{1}{2}$%.

If a machine is returned to us at any time before it is 20 years old, we can rebuild it to new standard at a cost which increases over a period of 10 years, approximately constantly from nothing to 70% of the cost of manufacturing a new machine. After that, the rebuild cost remains constant at 70% of the new cost. Thus the value to us of a machine is decreasing by 7% of its new cost each year, so our net income from it is $5\frac{1}{8}$% of its new value - the difference between the rental and the depreciation. Of course, since the machine depreciates, the rate of return as a % of its *current* value increases and averaged over the 8-year lease we find a rate of return on capital invested of roughly $7\frac{1}{2}$%.

All this ignores the effect of inflation, of course. We have not been unduly perturbed by this whilst inflation remained under 10%, but we are concerned that, with the higher inflation rates currently prevailing, the rental remains constant in money terms while the value to us of the machine increases due to inflation, thus counteracting the depreciation of the machine on which the increase in rate of return depends.

I wondered whether you could investigate whether higher inflation does significantly reduce our rate of return on an individual machine and, if possible, on our leasing business as a whole.

Memorandum 2

6.2 The Leather Working Machinery Group – general lines of donor's solution

This problem was originally passed by the English Glove Machinery Company, a producer of manufacturing machinery, to the Operations Research department of their parent company. The approach adopted by the OR specialists was to build up a mathematical model by stages. Firstly they modelled the return on the leasing business without the effects of inflation. When they were satisfied with that model they refined it to include inflation. Finally they used the model of the return on a single machine as a component part of a model of the return on the business as a whole.

(a) Single machine without inflation

The cost of a new machine was taken to be V_0. Since depreciation, measured by the cost to the company of refurbishing the machine to 'as new' condition, is 7% per annum, the value of the machine at time t was modelled by

$$V(t) = V_0(1 - 0.07t) \qquad\qquad t < 10$$
$$V(t) = 0.3V_0 \qquad\qquad t \geqslant 10.$$

The gross rental income, $G(t)$, is one eighth of the value of a new machine for the first eight years and, in the absence of inflation when the cost of an equivalent new machine remains constant, 75% of that thereafter. Thus $G(t)$ was expressed as

$$G(t) = 0.125V_0 \qquad\qquad t < 8$$
$$G(t) = 0.09375V_0 \qquad\qquad t \geqslant 8.$$

The net income, $N(t)$, was obtained by deducting the depreciation on the machine from the gross rental. Thus

$$N(t) = 0.055V_0 \qquad\qquad t < 8$$
$$N(t) = 0.02375V_0 \qquad\qquad 8 \leqslant t < 10$$
$$N(t) = 0.09375V_0 \qquad\qquad t \geqslant 10.$$

The rate of return, denoted by $R(t)$, on the investment is the net income divided by the residual value of the machine. Thus

$$R(t) = \frac{0.055}{1 - 0.07t} \qquad\qquad t < 8$$

$$R(t) = \frac{0.02375}{1 - 0.07t} \qquad\qquad 8 \leqslant t < 10$$

$$R(t) = \frac{0.09375}{0.3} \qquad\qquad t \geqslant 10.$$

The average value of the machine during the first eight years period is

$$0.125V_0 \int_0^8 (1 - 0.07t)\mathrm{d}t = 0.72V_0.$$

The mean rate of return is, therefore, $0.055V_0/0.72V_0$, which is approximately the 7.5% mentioned in Dixon's letter.

(b) Single machine with inflation

The effects of inflation could have been taken into account in a number of ways. The OR department chose to represent the devaluation of money as a continuous process using an exponential function. Thus at an inflation rate of $100a\%$, £1 at time 0 is worth £$\exp(-at)$ at time t or, equivalently, goods costing £1 at time 0 will cost £$\exp(at)$ at time t. The analysis outlined in the previous section must be modified as follows:

$$V(t) = V_0\exp(at)(1 - 0.07t) \qquad\qquad t < 10$$
$$V(t) = 0.3V_0\exp(at) \qquad\qquad t \geqslant 10.$$

The value of a machine, measured in £s, increases as the value of the £ falls due to inflation. In contrast the rental is constant in £ terms except that, at the review point after eight years, the rental is set to 75% of the rental for an equivalent new machine at that time so

$$G(t) = 0.125V_0 \qquad\qquad t < 8$$
$$G(t) = 0.09375V_0\exp(8a) \qquad\qquad t \geqslant 8.$$

The depreciation of a machine is 7% of its current value per annum so the net rental is described by

$$N(t) = 0.125V_0 - 0.07\exp(at)V_0 \qquad\qquad t < 8$$
$$N(t) = 0.09375V_0\exp(8a) - 0.07\exp(at)V_0 \qquad\qquad 8 \leqslant t < 10$$
$$N(t) = 0.09375V_0\exp(8a) \qquad\qquad t \geqslant 10.$$

The equivalent modified expression for the rate of return is

$$R(t) = \frac{0.125\exp(-at) - 0.07}{1 - 0.07t} \qquad\qquad t < 8$$

$$R(t) = \frac{0.09375\exp(a(8 - t)) - 0.07}{1 - 0.07t} \qquad\qquad 8 \leqslant t < 10$$

$$R(t) = \frac{0.09375\exp(a(8 - t))}{0.3} \qquad\qquad t \geqslant 10.$$

In figure 6.1 the rate of return as a function of time for various rates of inflation is plotted. It is evident that the rate of return is comparatively high on elderly machines but this represents a relatively small sum of money as the residual value of the machine is low.

The average rate of return over the first lease period may be calculated as

$$\int_0^8 N(t)\mathrm{d}t \Big/ \int_0^8 V(t)\mathrm{d}t$$

and has the values shown in table 6.1.

Table 6.1. *Average rate of return against inflation*

inflation (%)	5	10	15	20	25	30
average rate of return (%)	4.5	1.7	−0.8	−3.0	−4.9	−6.5

It is evident from that table that the profitability of leasing a machine was very much reduced even by inflation of 10% and that the levels of inflation prevailing at the time of the problem (between 20% and 30%) made the business unprofitable.

(c) *Leasing business as a whole*

At any time the company has a range of machines of different ages leased out to customers. The rate of return on machines of different ages is different. A model of the rate of return on the business as a whole needs some knowledge of the company's investment history. As a first model the OR department assumed that the investment history was constant in real terms, i.e. the company had manufactured and leased out a roughly equal number of machines of the same type in each past year. They also made the assumption that all machines were leased for a fixed period of 14 years (based on the comment in the memorandum from Dixon). The expressions previously

Fig. 6.1. Rates of return against time for various inflation rates.

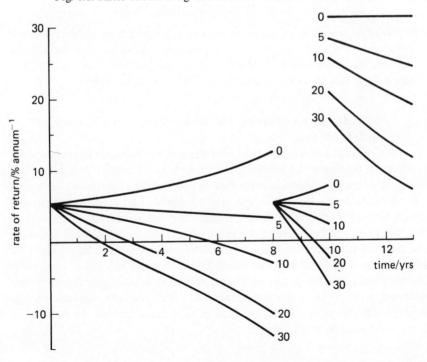

Table 6.2. *Rate of return on business as a whole against inflation*

inflation (%)	0	5	10	15	20	25	30
return on all business (%)	11.2	7.9	5.3	3.3	1.6	0.3	−0.8

derived for the value and the rate of return on machines as functions of their age could be used in this model with the modification that V_0, the value in £s of a machine at its time of manufacture, had to be replaced by $V_1 \exp(-at)$, where V_1 represented the value of a new machine at the time of modelling. Taking $n(t)$, the number of machines manufactured and leased t years ago, as a constant n_0 it was evident that the current value of the machines on lease is given by

$$n_0 V_1 \left\{ \int_0^{10} (1 - 0.07t)\mathrm{d}t + \int_{10}^{14} 0.3\mathrm{d}t \right\} = 7.7 n_0 V_1.$$

The net return on the machines is given by

$$n_0 V_1 \left\{ \int_0^8 (0.125\exp(-at) - 0.07)\mathrm{d}t \right.$$

$$\left. + \int_8^{10} (0.09375\exp(a(8-t)) - 0.07\mathrm{d}t + \int_{10}^{14} 0.09375\exp(a(8-t))\mathrm{d}t \right\}$$

$$= n_0 V_1 \{ [0.21875 - 0.125\exp(-8a) - 0.09375\exp(-6a)]/a - 0.7 \}.$$

The rate of return on the business as a whole calculated from these results has the values shown in table 6.2. From these results it was evident that the business as a whole was showing a decided loss of profitability by the time inflation had reached 10% and is almost certainly unviable for higher levels.

6.3 The Leather Working Machinery Group – commentary on student solutions

One of the interesting aspects of this study is the use of the exponential function to represent the effects of inflation. It is, of course, possible to use formulae of the compound interest type in place of the exponential functions. The choice between the two possibilities is probably best made on the basis of which method is more familiar to the modellers involved and which method they will therefore feel more comfortable with. If compound interest formulae are used the integrations used in the models will, of course, be replaced by summations. The results derived using the two approaches will differ slightly quantitatively but should exhibit reasonable qualitative agreement. There does not seem to be any strong reason for declaring one approach to be superior to the other. Most student groups have chosen to follow the compound interest formula path.

A second choice that must be made is whether to work in terms of money units, as the donor organisation chose to do, or in value units. When working in money units the value of an asset, expressed in money terms, increases as inflation takes its toll. If value units are chosen then values remain constant regardless of inflation but assets which are expressed in money units (e.g. the rental which is constant in money units) lose value as inflation acts. The two approaches are equivalent but represent different perspectives. Again there is no reason for declaring one to be superior to the other and the choice is probably best made on personal preference and prejudice.

In the author's experience this study usually provokes uncertainty and hence discussion on two main fronts. The first of these is the manner of presentation of the conclusions of the modelling. The second is the assumption regarding the investment history that needs to be made if a model of the whole business is to be derived.

The variation of rate of return with time shown in figure 6.1 applies if inflation is constant. In reality it is unlikely to remain constant over such long periods. A variation on such a diagram which was used by the donor of the problem and has also been used by several groups of students over the years is to assume increasing or decreasing inflation, i.e. instead of taking the inflation rate a as a constant using a relation such as

$$a(t) = a_0 + bt$$

where b may be a positive or negative constant. Provided the rate of change of inflation is not too violent such diagrams do not provide any surprises.

The assumption, for investment history, of a constant investment in new machinery and a fixed lease life of 14 years can also easily be relaxed. An investment history of the type

$$n(t) = n_0 \qquad\qquad\qquad\qquad\qquad\qquad t < t_1$$
$$n(t) = n_0(t_2 - t)/(t_2 - t_1) \qquad\qquad\qquad t_1 \leqslant t < t_2$$

may be substituted and the effects of using different values for t_1 and t_2 may be investigated. Some groups of students have tried out such history functions again without radically changing the conclusions.

Another point of interest is the range of values of inflation to be investigated. This problem originally arose during a period when inflation in Great Britain was running at previously unprecedented rates. The inflation rate, measured on a month to month basis, did, for a time, exceed 25%. The choice of the range of rates of inflation covered by figure 6.1 is obviously conditioned by these circumstances. It is noticeable that, in the middle and late 1980s, students rarely choose to consider inflation rates in excess of 15% unless prompted to do so. It is evident from the results shown in tables 6.1 and 6.2 that rates of inflation of 20% or more rendered the business totally unprofitable whilst at 5% to 10% profitability is reduced but not disastrously so. Whilst very high rates of inflation gave additional urgency to this problem, the effects of lower rates of inflation are still sufficiently striking to make the problem an interesting and challenging one for the students.

The final point which is brought out by this study concerns the reporting format. It seems fairly certain that the OR department's conclusions are damning for the current form of leasing contract when inflation of over 10% to 15% prevails. The obvious next question is to evaluate various alternatives. In the final analysis it must be the responsibility of Mr Dixon and the Commercial Department to define the terms of the lease in the light of their judgement of what will be acceptable in the market place. They need some guidance, however, about the magnitude of change needed to bring about a desired improvement in profitability. This guidance must avoid excessive prescription. The OR department of the donor organisation provided some projections of the effects of introducing an additional rental review at the 4 year stage and also of bringing the review time forward from 8 years to 6 years. It is appropriate to encourage students to do something similar, particularly as this will bring them to the point of using their mathematical judgement creatively in proposing possible formats for amended leasing contracts. It is probable that they will need considerable encouragement or even direction to do this. The author's experience is that students are very reluctant to go beyond what they perceive as the explicit brief of the problem. They are happiest confirming or refuting a specific proposal made in the source materials and markedly reluctant to be innovative and make suggestions not directly requested.

7

British Knitted Garments Group

7.1 British Knitted Garments Group – source documents

Memo

British Knitted Garments Group	*To* T. Fellowes
Subject	
'Press Offs' in Hosiery Manufacture	*From* C. Brown
	8 November

As you will see from the attached memos and report, there has been some concern in the Group about the effect of relative humidity on 'press off' rates in hosiery manufacture. The problem has been passed to us and I would like you to study the data and let me know your opinion of the importance or otherwise of the relative humidity in determining 'press off' rates. Could you also relate this to the overall efficiency of production and let me know what improvements we could expect from the installation of air-conditioning?

Memorandum 1

Memo

British Knitted Garments Group	To C. Brown Esq
	Head of Operations Research
Subject	
'Press Offs' in Hosiery Manufacture	*From* Dr A. Richards
	Research Department
	6 November

As a result of the decision of a recent monthly Research Liaison Meeting (see attached extract from Minutes), my Department has carried out a study of the effect of temperature and relative humidity on hosiery knitting yarn used by manufacturers in this Group. Laboratory tests have been devised to measure the tensile strength and resilience of the yarn under a number of loading conditions. The results indicate that the yarn properties are extremely insensitive to temperature changes under constant humidity conditions, but that the properties are somewhat sensitive to changes in relative humidity. The static tensile strength is relatively insensitive, but there is evidence that the strength of the yarn under rapidly changing tension may be degraded by high humidity. Because of the difficulty of measuring and reproducing the exact working conditions of the yarn on Gemini tight-knitting machines it was decided to carry out a programme of field observations at the Legley factory. A copy of the report of these field trials is attached.

You will see from the scatter diagram of the trials data that there is a somewhat confused relationship between 'press off' rate and relative humidity in practice. It is my feeling that your Department is probably better placed to undertake the detailed analysis of the data and I have obtained the approval of the Group Directorate to pass the problem over to you.

Memorandum 2

Extract from the Minutes of the Monthly Meeting of the Research
Department of the British Knitted Garments Group and
Representatives of the Manufacturing Subsidiaries.

25 September

===

Mr. King of Legley's, one of the hosiery manufacturing companies
within the British Knitted Group, raised the problem of the high
incidence of 'press off' faults on tight-knitting machines which occurs
during hot weather. These faults occur when the yarn used to
manufacture the tights breaks during knitting and results in the part-
knitted tights being discarded by an operative and the machine having to
be reset and restarted. Manufacturing efficiency is greatly reduced by
a high 'press off' rate. At Legley's the manufacturing efficiency,
which is computed weekly by comparing the actual number of tights
produced with the theoretical capacity, is reduced during the summer
months and Mr. King claimed that this was due to a higher 'press off'
rate. In informal contacts with other tight manufacturers in the Group
he had discovered that not all companies shared his experience of
reduced efficiency during the summer months and, furthermore, that those
which did not all had air-conditioned knitting floors. He thus
concluded that the increase in 'press offs' was a result of the higher
temperature or humidity during the summer, and requested that
consideration be given by the Group to fitting air-conditioning at
Legley's in order to reduce 'press offs' and restore manufacturing
efficiency during the summer months. His observations were supported by
representatives of several other hosiery manufacturers in the Group; in
particular, Mr. Neal of Knitights observed that very little difference
in efficiency could now be detected at their factory, which had been
fitted with air-conditioning just over a year ago.

It was agreed that the Research Department would investigate the
connection between temperature, humidity and 'press off' rate and
determine what improvement in efficiency might be expected from air-
conditioning. Discussions could then follow on the economic return from
the investment in air-conditioning. The plant at Knitights, a factory
of comparable size to Legley's, cost £143,000.

Attachment to memorandum 1

Brief Report No. BKD/RD/71/14	Circulation Standard _ _ _ _ _ _ _ _	_ _ _ _ _ _ _ _ _ _
Written by: B.J. Greenall	_ _ _ _ _ _ _ _ _ _ _	_ _ _ _ _ _ _ _ _ _
Approved by: A. Richards	_ _ _ _ _ _ _ _ _ _ _ _	Library (2) _ _ _ _ _

Title: Field Observations of Press Off Rates and Relative Humidity at
Legley's Hosiery Factory.

Objective:

To establish the connection or otherwise of press off rates
and relative humidity in hosiery manufacture.

Conclusions and Recommendations:

There appears to be some relation between press off rates
and relative humidity but further investigations must be
carried out before any recommendations can be made to
improve the efficiency of hosiery manufacture.

This Report is for the use of the recipient only.

Technical report, page 1

Introduction

A connection between the press off rate due to yarn breaks on Gemini tight knitting machines and temperature and/or humidity has been postulated by the tight manufacturing companies. Laboratory experiments were devised to test the hypothesis. These indicated that yarn strength was insensitive to temperature changes but was affected by changes in relative humidity (RH). The tensile strength under uniform loads showed little change, but the strength under conditions of rapidly changing load did vary considerably with RH (see Report No. BKG/RD/71/12). Such conditions of rapidly changing loads occur in parts of the tight knitting cycle. Due to the difficulty of reproducing, in the laboratory, the exact conditions of transient loading that occur, it was decided to carry out a series of controlled observations in a tight manufacturing factory.

The Observations

The trial was conducted over a period of $2\frac{1}{2}$ weeks at the Legley factory. During this period the RH on the knitting floor and details of each stop on 30 machines were recorded. The trial took place under normal production conditions and no attempt was made to deliberately vary ambient conditions. The knitting floor is divided into 'alleys', each consisting of 2 rows of 15 machines facing each other (see Figure 1). The knitting floor studied consisted of three alleys and one was chosen for the study. Each alley of machines is serviced by one operative who resets the machines following press offs, makes running repairs following certain other fault conditions (e.g. replaces broken needles), replaces packs of yarn as they run out and calls the mechanic if more serious faults develop. On the trial alley a record was made of the number of press off and other faults occurring in each 4-hour period and the time taken to repair each fault. A temperature/RH recorder is permanently installed on the floor (location 1 on Figure 1) and checks were made with a sling hygrometer at locations 2 to 8 during each 4-hour period to determine the variation of temperature and RH over the knitting floor.

The Data

The correlation between the air conditions recorded at locations 1 to 8 was extremely high ($\pm 1^{0}$C, $\pm 2\%$ RH) and no systematic variation was detected. It was concluded that the conditions at location 1, the permanent recorder, could be taken as typical of the floor as a whole. Table 1 shows the recorded

Technical report, page 2

(2)

press off rate for each 4-hour period, expressed in terms of press offs per
30 machines per 8-hour shift, and the mean RH for each period, measured by
the fixed recorder at location 1. The mean number of faults other than
press offs was 62/alley/shift. The average time taken to repair a fault of
any kind was 2.7 minutes. Table 2 is an extract from the factory records
provided by the Manager. This was part of the evidence on which he based
his original assertion that RH affects efficiency. In this case the RH and
efficiencies are averages over 1-week periods.

Conclusions

Figure 2 shows a scatter diagram of the data recorded in Table 1. A
rather confused trend is discernible with press offs/shift rising somewhat
as the RH increases. The data in Table 2 is also plotted on a scatter diagram
in Figure 3, and again shows a distinct fall in efficiency as RH increases.
It is recommended that the data be further analysed to determine the magnitude
of the expected change in press off rate as humidity changes and to relate
this to efficiency.

Technical report, page 3

Results of Trials for Press Off Rates/Shift/Alley Against Mean RH
Measured Over 4-Hour Periods

Period	Press Offs	RH	Period	Press Offs	RH
1	40	35	34	48	15
2	70	32	35	60	15
3	52	28	36	48	15
4	66	27	37	23	16
5	14	26	38	23	16
6	22	25	39	23	16
7	38	24	40	29	17
8	44	24	41	48	20
9	34	26	42	52	25
10	42	30	43	21	32
11	28	33	44	27	34
12	18	28	45	6	35
13	38	24	46	31	35
14	38	22	47	35	36
15	18	19	48	35	36
16	34	19	49	52	36
17	14	19	50	12	36
18	12	16	51	12	34
19	40	16	52	41	32
20	40	16	53	116	40
21	30	17	54	84	39
22	26	17	55	76	39
23	22	17	56	71	35
24	33	16	57	94	35
25	33	15	58	87	39
26	10	16	59	66	38
27	12	16	60	45	32
28	27	18	61	72	36
29	25	19	62	72	37
30	29	19	63	54	35
31	29	19	64	43	37
32	12	18	65	41	36
33	19	16			

TABLE 1

Technical report, page 4

Extract from Records of Efficiency v Mean RH for One Week Periods (August to February)

Week Ending	Efficiency	Mean RH
8 August	86%	40
15 August	88%	38
22 August	93%	37
29 August	90%	39
5 September	84%	40
12 September	93%	39
19 September	95%	32
26 September	92%	36
3 October	94%	33
10 October	93%	35
17 October	95%	34
24 October	94%	30
31 October	97%	30
7 November	96%	28
14 November	95%	24
21 November	95%	26
28 November	94%	27
5 December	96%	23
12 December	96%	26
19 December	95%	21
26 December *	91%	18
2 January *	88%	25
9 January	94%	16
16 January	96%	19
23 January	92%	22
30 January	95%	15
6 February	90%	31
13 February	98%	24
20 February	96%	20
27 February	98%	17

* Data based on incomplete week's operations

TABLE 2

Technical report, page 5

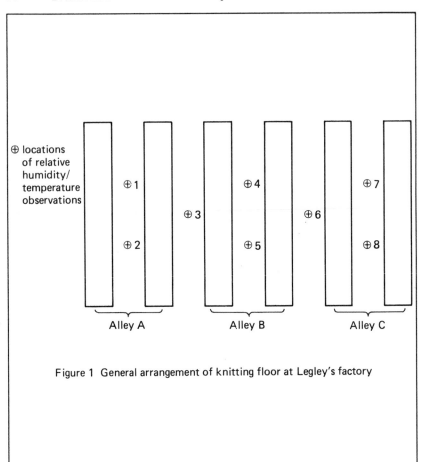

Figure 1 General arrangement of knitting floor at Legley's factory

Technical report, page 6

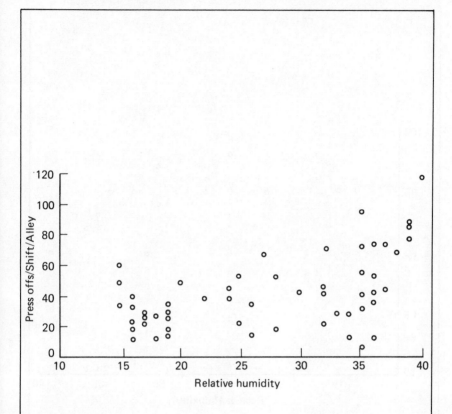

Figure 2 Scatter diagram of observations of press offs and relative humidity

Technical report, page 7

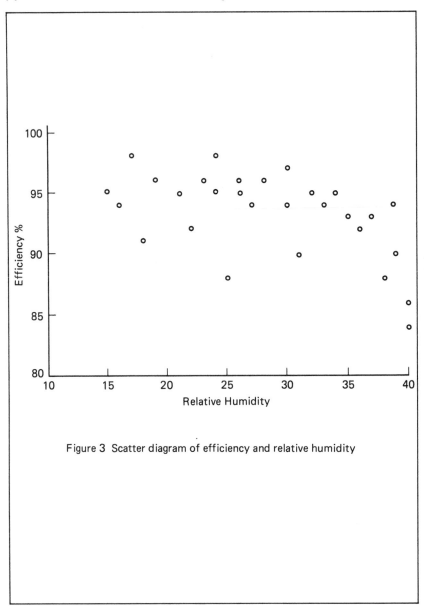

Figure 3 Scatter diagram of efficiency and relative humidity

Technical report, page 8

7.2 British Knitted Garments Group – general lines of donor's solution

The data for press offs as a function of relative humidity presented in table 1 and plotted in figure 2 of the report showed substantial scatter. The Operations Research team who faced this problem decided to adopt a curve fitting technique to try to discern the underlying pattern in the scatter. A quadratic curve was fitted to the data by the least squares technique and resulted in the curve

$$po = 0.145rh^2 - 6.358rh + 95.690.$$

This curve is shown superimposed on the press offs/relative humidity diagram (figure 2 of the report) in figure 7.1.

As a check on this relation they then tried to relate this to the efficiency observed. The report states that the mean number of faults other than press offs observed was 62/alley/shift and that the mean repair time for any fault, including press offs, was 2.7 minutes. A model of the efficiency to be expected at a given level of press offs was taken to be

$$\text{efficiency} = \frac{100\,(\text{total time} - \text{lost time})}{\text{total time}}\%$$

$$= \frac{100(8 \times 60 \times 30 - 2.7(62 + po))}{8 \times 60 \times 30}\%$$

(assuming 8 hour shifts and 30 machines per alley)

i.e. efficiency $= 97.04 + 0.119rh - 0.00272rh^2$ %.

This curve is shown superimposed on the efficiency/relative humidity

Fig. 7.1. Scatter diagram of press offs/relative humidity with quadratic and linear least squares curves fitted.

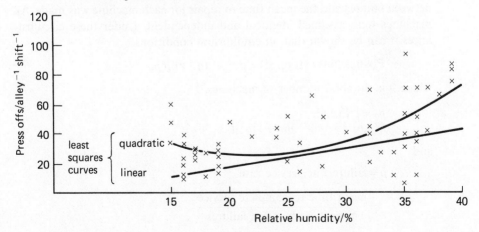

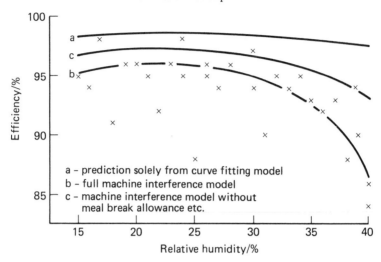

Fig. 7.2. Scatter diagram of efficiency/relative humidity with predicted curves from various models.

diagram (figure 3 of the report) in figure 7.2. As a result of this check, which indicated that fault rate alone does not predict the efficiency of the plant well, they were prompted to seek other explanations.

The explanation they arrived at involved machine interference theory, a special case of queueing theory in which the queuers are not drawn from an infinite population but from a finite one. As a result the probability per unit time of new members joining the queue falls as the size of the queue grows until, when the whole population is in the queue, the probability of new members joining becomes zero. The machine interference problem is dealt with by many texts on queueing theory (e.g. Cox and Smith, 1961) or may readily be derived from first principles.

The assumption of negative exponential distributions for the mean time between failures and the mean time to repair for each machine was made. All machines were assumed identical and independent. Under these circumstances it can be shown that, in equilibrium conditions,

$$P(r \text{ machines stopped}) = p_r = n!/(n-r)!\rho^r p_0$$

where n is the total number of machines,

$$p_0 = 1/F(\rho,n),$$
$$F(\rho,n) = 1 + n\rho + n(n-1)\rho^2 + \ldots + n!\rho^n,$$

and

$$\rho = \text{failure rate/service rate}$$

$$= \frac{\text{mean time to complete service}}{\text{mean time between failures}},$$

Table 7.1. *Efficiency versus relative humidity*

relative humidity	15	20	25	30	35	40
efficiency (%)	95.3	95.9	95.9	95.0	92.7	86.4

(see, for instance, Cox and Smith, op. cit.). Hence the average number of machines stopped is $\Sigma r p_r$ and efficiency is $100(1 - \text{mean number stopped}/n)$.

Two complicating factors were taken into account. Firstly the data from which the failure rates were derived were, strictly speaking, expressed in failures per machine running hour. Hence

$$\text{failure rate} = \frac{\text{press offs/shift}}{8 \times 60 \times 30 \times E}$$

(assuming 30 machines per alley and an 8 hour shift) where E is efficiency. There was thus a circular situation. The ratio ρ depended on E and E, in its turn, depended on ρ. Fortunately, if E was assumed to be 1 and a value for ρ found, then that value used to find a new value of E and the iteration repeated until E changed negligibly, the convergence was found to be rapid. The second complication was that the operatives servicing the machines were not actually available for the whole of an 8 hour shift. Meal breaks and time out for calls of nature were assumed to reduce the availability of operatives to 7 hours per shift. This was approximately accounted for by increasing the mean service time in proportion to the non-availability of the operatives. Thus

$$\text{service rate} = \frac{\text{available minutes}}{\text{mean repair time} \times 8 \times 60}$$

$$= 0.324 \text{ services/min.}$$

Table 7.1 shows the results of this calculation and these values are also plotted on figure 7.2. It is evident that, with machine interference accounted, the fall in efficiency with rising relative humidity is much better explained.

The original task was to determine the effect of relative humidity on press off rate and to quantify, if possible, the improvement in efficiency that could be expected if air-conditioning were installed in the Legley factory. Whilst it could not be denied that controlling the relative humidity in the factory would improve the efficiency, as claimed by the Legley factory manager, the insight provided by the machine interference analysis allowed the OR team to make an alternative suggestion.

The degradation in efficiency was partly due to the amount of time machines spend stopped and awaiting attention from an operative. If such waiting time could be reduced the efficiency would be improved. One way of doing this would be, instead of assigning one operative to each alley of 30 machines, to assign all three operatives to look after all the 90 machines on the knitting floor

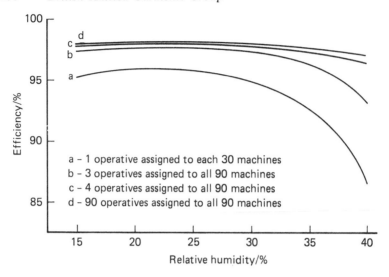

Fig. 7.3. Efficiency/relative humidity curves predicted by machine inter-ference model for various conditions.

regardless of their location. This corresponds to a multiple server queue and results for this situation are again available in the literature (see, for instance, Gross and Harris, 1974). For a queue with m servers the result is

$$p_r = \frac{n!}{r!(n-r)!}\rho^r p_0 \qquad\qquad 0 \leqslant r < m$$

$$p_r = \frac{n!}{(n-r)!m!m^{r-m}}\rho^r p_0 \qquad\qquad m \leqslant r \leqslant n$$

where

$$\frac{1}{p_0} = \sum_{r=0}^{m-1} \frac{n!}{r!(n-r)!}\rho^r + \sum_{r=m}^{n} \frac{n!}{(n-r)!m!m^{r-m}}\rho^r.$$

The efficiency to be expected as a function of the relative humidity was found for this case as before. Figure 7.3 shows curves for efficiency if each operative looks after his or her own 30 machines, for 3 operatives looking after 90 machines, for 4 operatives looking after 90 machines and for 90 operatives looking after 90 machines. The last case provided the comparison with the best that could be achieved since, with one operative per machine, it would be guaranteed that no machine ever had to wait for service.

In practice constraints of the geography of the factory floor and the distances operatives would have to move to service remoter machines would probably mean that the full potential improvements in efficiency would not be realised but these figures gave an indication of the potential for improvement.

The possibility of employing seasonal labour (such as university students on vacation) to augment the regular staff during periods when high humidity is normal might be considered. In this way the benefits of the improved efficiency apparent when using 4 operatives during periods of high humidity might be available without the cost of employing the extra staff permanently.

As a result of this work the OR department were able to provide the management with not only a reasonable estimate of the improvement in efficiency that could be expected from various standards of air conditioning but also an alternative possible way of alleviating the efficiency problem, that is reorganisation of the working practices of the repair operatives and the possible employment of one or more additional operatives. The decision as to which option, if any, to implement would then be taken in the light of detailed costings of the options and financial benefits.

7.3 British Knitted Garments Group – commentary on student solutions

This is an interesting case study because the original problem posed was not the main problem, or at least not the whole of it. In the real world mathematical modellers may often be presented with problems by people who already know what they want the answer to be and are really only asking for some justification of their solution. In extreme cases the mode of presentation of the problem and the data may be slanted in such a way as to attempt to force the mathematician's hand. Modellers have to be aware of the possible existence of such non-mathematical factors. In this case the factory manager was probably unaware of the importance of the machine interference effects so any obstruction of the investigation was accidental rather than deliberate but the effect was still to obscure the problem.

The first part of the problem is to discern any underlying pattern in the relation of press offs and relative humidity. Visual inspection of the scatter diagram, figure 2 of the report, suggests that there is a relation with press offs rising at high humidity and, to a lesser extent, at low humidity and having a minimum around an RH of 20. A considerable amount of scatter around this relation is apparent. There are two immediate options for fitting a curve to represent this relationship. Either a line may be sketched by eye or some mathematical technique may be used to fit a curve. The extent of the scatter makes the first option a rather dubious process. An appropriate mathematical way of obtaining a curve representing this relation is through least squares curve fitting. Student groups have chosen to fit both quadratic curves and straight lines to the data. The quadratic curve is given above whilst the straight line fit is

$$po = 1.335rh + 4.598.$$

The quadratic curve would probably be deemed the more appropriate by most observers though an argument can certainly be made for the straight line curve fit. It should also be noted that the matrix equation used to derive the coefficients of the least squares curve is extremely ill-conditioned. A 1% variation of any of the coefficients in the matrix of summations for the quadratic curve fit results in variations in the coefficients of the curve of, in the worst cases, 100% or more. The variations resulting in the straight line case also exceed 70% in the worst cases. Some variation in the results produced by students is therefore to be expected. The results quoted here have been obtained on a micro-computer which uses 40 bit real number arithmetic. Some students, in the author's experience, had some trouble in obtaining curves which they could believe when plotted back on the scatter diagram. This was, in some cases, caused by over enthusiastic rounding of the coefficients in the matrix of summations.

Students do not necessarily try to extrapolate from the dependence of press offs on relative humidity to the relation of efficiency to relative humidity. It is more often than not necessary to prompt them in some way to make the connection. When they make the attempt they find, as did the OR department of British Knitted Garments, that the press-off rate cannot totally explain the loss of efficiency. In the author's experience a little discussion of the problem usually results in the students realising that random breakdowns will almost certainly result in machine interference effects. Once that realisation is achieved they can be asked to investigate, probably using the library, whether anything is known about such problems and then to apply the results of that research to this problem.

The author, if asked for suggestions about suitable books, offers a list of books on queueing theory for students to investigate. The list quite deliberately includes some books which contain no directly relevant inform-ation and some that are not available in the local faculty library. This is a reasonable simulation of the help that they would probably receive from a section leader or departmental head in industry; such a person would have greater experience in the field and so might well be familiar with a range of books that might have bearing on the problem but would be less likely to be able to identify immediately a single book guaranteed to contain the solution.

The machine interference calculation carried out by the BKG OR team included attempts to account for the two complicating factors described in the previous section. If these two factors are ignored predictions of efficiency midway between the results obtained without accounting for interference and with all factors accounted are obtained. These results are also plotted on figure 7.2. It is evident that these nearly account for the observed efficiency though the predictions are at the upper bound of the scatter range.

The final stage of the BKG team's investigation, the analysis of possible alternative working patterns, is an interesting one for students. In the author's experience if this problem is given to a student group fairly early in the course they are unlikely to respond very readily to the open-ended challenge of suggesting alternative working practices. On the other hand, if the problem is given to a group who have worked on a number of case studies already and who are therefore more familiar with the demands of this type of problem they are much more likely to respond with creative ideas and mathematical investigation to match. There is little that can be said about the range of suggestions that might be made other than that the tutor must judge their good sense and viability and comment appropriately.

8

Gamma Avionics

8.1 Gamma Avionics – source documents

internal memo	project
	M6 RADAR
gamma a.s.	
To T. Fellows Mathematical and Computational Services	From A. Williams M6 Project Manager 6th June

Subject: Antenna drive motor power requirement for M6b nodding radar

The antenna of the M6b nodding radar is driven by a long link
(A on the attached drawing) connecting the antenna and a crank on a
driven shaft (B). The driven shaft is mounted in bearings (C) on a
swinging arms (D) which are located by screw jacks (E). Thus the nodding
angle is mainly determined by the length of the crank, and the nodding
sector (i.e. the mean orientation of the antenna) is controlled by the
position of the swinging arms which is adjusted by the screw jacks. The
driven shaft is powered by a geared constant speed motor (F) and nominally
rotates at 1 Hz. This speed must be accurately maintained and, in order
to choose a motor capable of maintaining speed, the maximum torque
required to drive the antenna must be known. The antenna pivot axis (G)
is chosen so that the unit is statically balanced, so the motor has to
overcome dynamic loads only. A simple calculation carried out by Bob
Inglis, attached to this memo, provides an estimate of the maximum torque
based on the assumption of simple harmonic motion throughout. Obviously
the geometry of the system is such that the true motion will deviate from
SHM but the magnitude of such deviation is uncertain. Could you please
investigate the true motion of the system and determine the limits of the
torque that must be provided by the motor to maintain motion over the whole
range of allowable positions of the swinging arms.

Memorandum 1, page 1

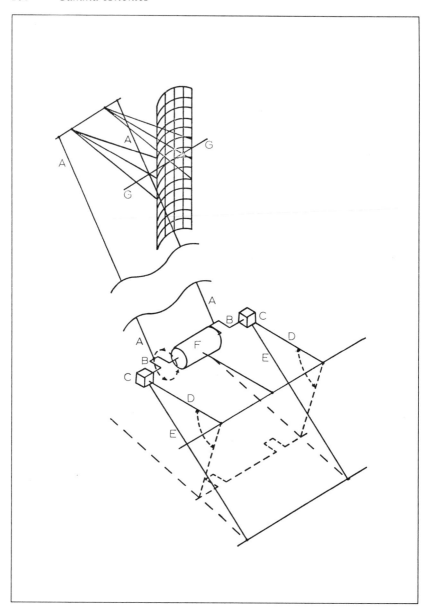

Memorandum 1, page 2

internal memo	project
gamma a.s.	M6 RADAR
To Mr. A. Williams M6 Project Manager	From R. Inglis 5th June

Subject: Estimation of power requirement for drive motor (M6 nodding radar).

    The drive linkage for the nodding antenna of the M6b radar can be represented as shown in the sketch attached. As the crank OC rotates about O the link BC drives the antenna AB through approximately $\pm 9^0$. Since the distances AD and BC are relatively long the motion of B is approximately SHM of amplitude 2.5". The crank rotates at 1Hz = 2π rad/s so we may put $\phi = 2\pi t$ and thus

$$\theta \simeq \theta_0 + \frac{2.5 \cos 2\pi t}{16}$$

where θ_0 is determined by the angle, α, of the swinging arm OD and the line AD.

    Since the antenna is statically balanced, its equation of motion can be given as

$$I\ddot{\theta} \simeq 16F$$

where I is the moment of inertia of the antenna about A and F the compression in the link BC.

    The torque, T, required of the motor is given by

$$T \simeq -F \times 2.5 \sin \phi$$

so, eliminating F and substituting for $\ddot{\theta}$ and ϕ, we have

$$T \simeq 2.5 \sin 2\pi t \ \frac{I}{16} \frac{2.5}{16} (2\pi)^2 \cos 2\pi t$$

i.e.

$$\frac{T}{I} \simeq .964 \sin 2\pi t \cos 2\pi t = .482 \sin 4\pi t$$

or

$$\frac{T}{I} \simeq .482 \sin 2\phi$$

Thus the torque required to drive the antenna is a sinusoidally varying quantity with maxima when $\phi = 45^0$ and 225^0 and minima when $\phi = 135^0$ and 315^0. At the minima the motor is, in fact, retarding the antenna against its own rotational inertia. The maximum torque required of the motor is, therefore, .482 I plus an allowance for the frictional losses in the linkage.

    This analysis is only approximate and may be inadequate when α, and therefore θ_0, differ appreciable from 90^0. The operation of the radar calls for α to vary between the limits 50^0 and 130^0 so the effect of this variation should be investigated.

Memorandum 2, page 1

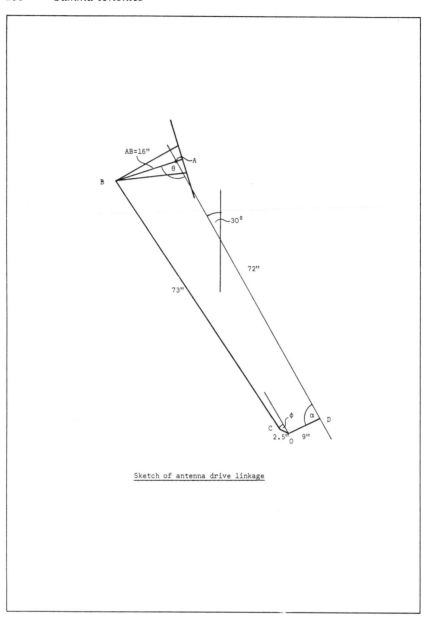

Sketch of antenna drive linkage

Memorandum 2, page 2

gamma

avionic systems

THE PRINCIPLES OF NODDING RADAR

A nodding radar consists of an antenna mounted on a horizontal axis with a drive mechanism causing the antenna, through a system of cranks, to oscillate over a small range about some mean elevation. The whole subsystem is mounted on a carriage pivoted about a vertical axis and independently driven so that the nodding antenna can be oriented in any desired direction. The system can be used as an independent radar tracking system, providing full range, bearing and height information on any target within its range. Used in this way however the system has a relatively low target acquisition rate.

Nodding radar is commonly used as one element of a two element complete radar system. A conventional fan beam radar provides range and bearing data. The nodding radar is slaved to this main (or surveillance) radar and provides height information on any target observed by the main radar. The main radar rotates continuously about a vertical axis and radiates a broad fan shaped beam in a vertical plane, thus providing maximum height and range coverage. Reflections are received as this beam passes the target bearing. The main radar commands the nodding radar to home on the target bearing and the nodding radar then sweeps a sector of sky in a vertical oscillatory manner. Its beam is also fan shaped but in a plane perpendicular to that defined by the target position and the vertical axis through the radar position. Thus it can accurately determine the target's elevation above the horizontal, and this information combines with the range information to yield the target's height. An oscillating, as opposed to a rotating, scan is used since the range of possible target elevations is confined to a small sector of sky. In its simplest form the antenna nodding mechanism can be driven by a simple crank and link giving a predetermined nodding angle. Usually, however, some method of altering geometry is incorporated so that the mean angle of elevation of the nodding antenna can also be changed. This allows the nodding angle to be fairly small (typically $\pm$ 10^0) and so optimises the data acquisition rate of the system.

Nodding radar is one of the many Gamma Avionic developments which contribute to the safety of airline operations all over the world.

This information is released by the Sales Division of Gamma Avionics who will be pleased to provide further information on any of the company's products.

Background document

8.2 Gamma Avionics – general lines of donor's solution

The mathematicians at Gamma Avionics identified the drive mechanism for the nodding radar as essentially a form of five bar linkage as illustrated in figure 8.1. Using the notation of that figure the mechanism may be described as follows. The link AB carries the motor at B and is supported by the screw jacks. The angle α is, therefore, constant during the primary motion of the system, but may be altered between values of $50°$ and $130°$. The motor drives the crank BC which, through the link CD, causes the arm DE, carrying the radar antenna, to rock up and down about the pivot E. The angular velocity of the link BC must be maintained constant so the constant speed motor chosen to drive the system must have a rating sufficient to provide the required torque without significant speed deviation.

The system was analysed in two stages. Firstly, the torque required was related to the moment of inertia and the angular acceleration of the radar head. This indicated a need for knowledge of that angular acceleration. Therefore, secondly, relationships between the angles ψ and ϕ with α as a parameter and between θ and ϕ with α as a parameter were derived. A technique for solving these to obtain ψ and θ in terms of ϕ was developed. The second relationship was differentiated twice (once analytically and once numerically) to obtain expressions for $\dot{\theta}$ and $\ddot{\theta}$ in terms of the angles. Finally the results of both parts of the analysis were combined to obtain the required torque.

Since it was dynamically balanced the equation of angular motion of the radar antenna yielded

$$I\ddot{\theta} = F\sin(\theta + \psi)b$$

where I is the moment of inertia of the antenna about the pivot point E. Taking T as the torque developed by the motor the equation

$$T = F\sin(\phi - \psi)e$$

was obtained. Hence

$$\frac{T}{I} = \frac{\sin(\phi - \psi)e}{\sin(\theta + \psi)b}\ddot{\theta}. \tag{8.1}$$

This posed the problem of obtaining expressions for the angles ψ and θ and the angular acceleration $\ddot{\theta}$ in terms of ϕ. The required relations were derived from the physical constraint equations for the mechanism. Resolving the five bar linkage parallel to and perpendicular to the link AE resulted in

$$d\cos(\alpha) + e\cos(\phi) + a\cos(\psi) + b\cos(\theta) = c \tag{8.2a}$$
$$d\sin(\alpha) + e\sin(\phi) + a\sin(\psi) - b\sin(\theta) = 0. \tag{8.2b}$$

Eliminating θ from these two equations resulted in an equation of the form

$$k_1\cos(\psi) + k_2\sin(\psi) + k_3 = 0 \tag{8.3}$$

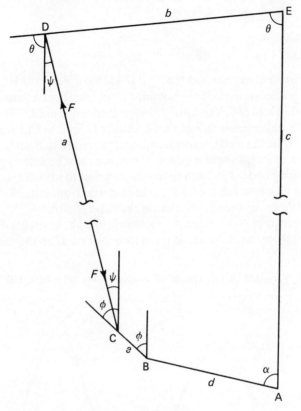

Fig. 8.1. Five bar linkage and notation.

where

$$k_1 = 2a(d\cos(\alpha) + e\cos(\phi) - c)$$
$$k_2 = 2a(d\sin(\alpha) + e\sin(\phi))$$
$$k_3 = a^2 - b^2 + c^2 + d^2 + e^2 + 2ed\cos(\phi - \alpha) - 2c(d\cos(\alpha) + e\cos(\phi)).$$

Manipulation of equation 8.3 resulted in

$$(k_1^2 + k_2^2)\cos^2(\psi) + 2k_1k_3\cos(\psi) + k_3^2 - k_2^2 = 0.$$

This has two roots for $\cos(\psi)$ but the appropriate one was easily identifiable. Having solved this to obtain ψ in terms of ϕ and α, θ was obtained from the relation

$$\tan(\theta) = \frac{a\sin(\psi) + d\sin(\alpha) + e\sin(\phi)}{c - a\cos(\psi) - d\cos(\alpha) - e\cos(\phi)}.$$

To obtain the angular acceleration equations 8.2 were differentiated yielding

$$e\sin(\phi)\dot{\phi} + a\sin(\psi)\dot{\psi} + b\sin(\theta)\dot{\theta} = 0$$
$$e\cos(\phi)\dot{\phi} + a\cos(\psi)\dot{\psi} - b\cos(\theta)\dot{\theta} = 0.$$

Elimination of $\dot{\psi}$ from these two gave

$$\dot{\theta} = \frac{e\sin(\psi - \phi)}{b\sin(\psi + \theta)}\dot{\phi}. \tag{8.4a}$$

The operational requirement was that $\dot{\phi}$ should be constant at 1 Hz, that is 2π rads^{-1}. The acceleration was then obtained from numerical differentiation of closely spaced values of $\dot{\theta}$. Values of T/I were thence obtained. The torque has both positive and negative values, that is at some points the function of the motor is to accelerate the mechanism and at others to retard it. Both functions place a load on a constant speed motor so the motor must be able to provide both an accelerating and a decelerating torque sufficient to maintain constant speed. Thus the absolute value of T/I is the important quantity. This goes through four maxima in each revolution of the drive motor.

A computer program was written to implement the above model. Figures 8.2(a) and 8.2(b) show the cyclic variation of torque demand on the motor. In

Fig. 8.2. Variation of T/I with ϕ (a) $\alpha = 50$, 70, 90; (b) $\alpha = 90$, 110, 130.

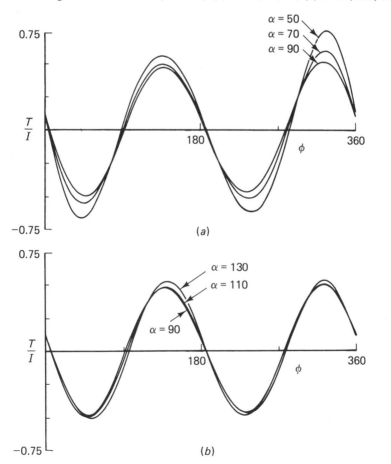

Table 8.1. *Maximum values of torque for various values of alpha*

α					
50	T/I	0.669	0.563	0.623	0.757
	ϕ	43	137	239	326
60	T/I	0.605	0.525	0.567	0.670
	ϕ	44	137	237	325
70	T/I	0.557	0.497	0.525	0.604
	ϕ	45	138	236	324
80	T/I	0.523	0.479	0.495	0.557
	ϕ	46	138	234	324
90	T/I	0.502	0.470	0.476	0.524
	ϕ	47	139	233	323
100	T/I	0.492	0.471	0.468	0.504
	ϕ	48	139	232	323
110	T/I	0.492	0.481	0.468	0.495
	ϕ	49	140	231	323
120	T/I	0.499	0.499	0.477	0.494
	ϕ	50	141	230	323
130	T/I	0.512	0.525	0.493	0.499
	ϕ	52	142	230	323

table 8.1 the values of the four maxima of absolute value of torque in each revolution with the corresponding values of ϕ are recorded for a range of values of α. It is evident that the worst case occurs at an angle α of 50° and is about 57% in excess of the predictions made, using the simplified analysis, by Inglis and described in his memo. Evidently the motor must be more highly rated.

8.3 Gamma Avionics – commentary on student solutions

This is a relatively closed problem. There is not a great deal of scope for variation from the donor's solution. One of the obvious variations, however, is in the derivation of $\ddot{\theta}$. It is unnecessary to resort to numerical differentiation. An equation parallel to equation (8.4a) can be derived showing that

$$\dot{\psi} = -\frac{e\sin(\theta+\phi)}{a\sin(\theta+\psi)}\dot{\phi}. \tag{8.4b}$$

Differentiation of equation (8.4a), noting that $\ddot{\phi}=0$, leads to

$$\ddot{\theta} = -\frac{e\cos(\phi-\psi)\dot{\phi}(\dot{\phi}-\dot{\psi})+b\cos(\theta+\psi)\dot{\theta}(\dot{\theta}+\dot{\psi})}{b\sin(\theta+\psi)}$$

This is readily evaluated using equations (8.4a) and (8.4b). The results of doing this agree, as would be expected, with those of table 8.1.

The problem demonstrates two important lessons for students. Firstly much use is made of implicit equations. The equations (8.2) which are differentiated to obtain $\dot\theta$ and $\dot\psi$ are not readily expressed in explicit form. Implicit differentiation must be used. In the author's experience students do not find this easy or obvious. The exercise therefore provides much needed practice in the use of such methods.

Secondly it is a good illustration of the principle of worst case analysis. The motor must be rated to provide the necessary torque in all cases. Its rating will therefore be determined by the maximum demand that is placed upon it for any value of α. This evidently occurs at 50°.

Some groups of students have attempted various approximations inter-mediate in accuracy between the analysis presented above and the simplified analysis of Inglis described in the memo. The author has yet to find any such approximate analysis a convincing alternative. The method adopted by the donor organisation is reasonably straightforward and not sufficiently time consuming for any alternative simpler analysis to offer convincing economies to offset the loss of accuracy and the inevitable uncertainty about the accuracy of the result.

Fig. 8.3. The equivalent four bar linkage for the approximate analysis.

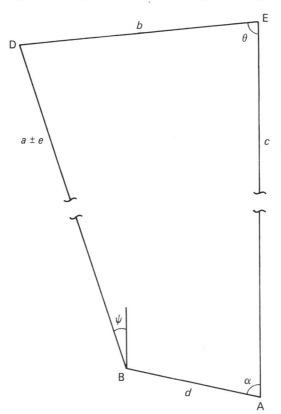

Table 8.2. *Ranges between extreme values of θ for various α.*

alpha	range
50	20.756
60	19.799
70	19.042
80	18.494
90	18.149
100	17.992
110	18.007
120	18.173
130	18.459

An example of one of these intermediate analyses is as follows. The extreme values of θ occur when the link BC is in line with the link CD (notation as figure 8.1). The distance between points B and D is then $a+e$, in which case angles ϕ and ψ are equal, or $a-e$, in which case $\phi=\psi+\pi$. The mechanism is, when in one or other of these extreme configurations, essentially equivalent to the four bar linkage shown in figure 8.3. Resolving this system parallel to and perpendicularly to AE gave the constraints

$$d\sin(\alpha) + (a \pm e)\sin(\psi) - b\sin(\theta) = 0$$

and

$$d\cos(\alpha) + (a \pm e)\cos(\psi) + b\cos(\theta) = c.$$

Eliminating ψ from these two equations gives

$$b^2 + c^2 + d^2 - (a \pm e)^2 - 2(bc\cos(\theta) + cd\cos(\alpha) - bd\cos(\theta+\alpha)) = 0.$$

This implicit equation was solved by Newton–Raphson iteration to yield the range between the extreme values of θ for different values of α. The results are shown in table 8.2. The ranges were then used to refine the results of the Inglis model. Instead of the relation

$$\theta = \theta_0 + 2.5\cos(2\pi t)/16$$

used by Inglis the student group substituted

$$\theta = (\theta_{max} - \theta_{min})\cos(2\pi t)/2.$$

The result of this modification is to increase the maximum predicted value of T/I by 16% to 0.559 occurring at $\alpha = 50°$. Whilst this is correct in predicting that the smallest value of α gives the worst case it considerably underestimates the magnitude of the worst case value of T/I. It is also difficult, if not impossible, to give any estimate of the accuracy of this improved approximation. For these reasons the author did not regard this as an appropriate alternative approximation.

Student groups who have successfully solved this problem often think that their results are incorrect because of the relatively large discrepancy between their own results and those from the simplified analysis by Inglis. Most do not, intuitively, expect such a large discrepancy and consequently reject their own results on the grounds 'There must be something wrong – the answer's three times as big as expected'. Another important lesson from this problem is therefore not to summarily reject the unexpected.

It is worth pointing out that the radar system which provides the basis of the problem posed in this chapter is a relatively elderly one. The range of options available today for achieving similar ends is considerably wider so a radar designed, as the nodding radar was, for searching a narrow sector of sky would not necessarily be designed on the same mechanical principle.

9

Fanning Industries

9.1 Fanning Industries – source documents

FANNING INDUSTRIES LTD

---

to   T. Fellows   **from**   F. James
                                       Head of Analysis Section

                              date   7th April

Large End Bearing Erosion on PS9 Series Diesels

As you will see from the attached memo, some problems
are being experienced with our new engine.   I'd like you
to undertake the analysis Mr. Grace has requested.   I've
obtained some sketches of the crankshaft/connecting rod/
piston assembly from the Drawing Office, and also asked
them to list the dimensions that are likely to be necessary
for your work.

I would like a verbal progress report from you next week,
and we will discuss - in the light of this - what further
analysis is possible, and any recommendations we can make.

Memorandum 1

FANNING INDUSTRIES LTD

to   F. James   **from**   M. Grace
         Head of Analysis Section   Chief Development Engineer

                                      date   6th April

Large End Bearing Erosion on PS9 Series Diesels

I attach a copy of a memo from Peter Thirsk concerning damage to large end bearings on the new PS9 series of large industrial static diesel engines. I have obtained samples of lubricating oil from both engines concerned and submitted them to our Metallurgy Department for analysis. They report that there is no undue contamination, and feel that the cause of the damage cannot be put down to foreign matter in the lubricant or to lubricant degradation.

Another possible explanation of the damage is that excessive oil flows at parts of the engine cycle are eroding the bearing shell. As with most engines of this type, the lubricating oil functions not only as lubricant but also as internal coolant. In fact, the mass flow rates of oil round the engine are determined by this latter requirement and are far in excess of that needed for lubrication alone. The piston crowns are cooled by oil which is fed from the crank case gallery, through drillings, into the main bearings and thence into drillings within the crankshaft. The crankshaft drillings lead oil into the big end bearings and so to drillings in the connecting rod up which the oil passes to reservoirs in the piston crowns. Oil spills from these reservoirs into the crank case sump and, via an oil cooler, back to the pump. It is the flow of oil in the connecting rod drillings which now concerns me. The PS9 series is a relatively high-speed engine for its size, and so the inertia forces on the oil in the connecting rod drillings as the engine operates must be relatively high. As the engine passes bottom centre, these forces must be acting against the oil pump pressure head and, if large enough, could cause oil to flow down the drilling and back into the large end bearing shell- the reverse of the designed flow direction. If this happened the erosion damage would not be unexpected. Similarly, as the engine passes top centre the inertia force on the oil in the connecting rod drilling will be enhancing the pump pressure head locally, and - should the mass flow in this drilling become excessive - the large end bearing could be starved of oil and cavitation result. Damage similar to erosion damage would be expected in this instance also.

Could your department please investigate the sources of inertia forces in the oil ways and give us some guidance as to the magnitude of these forces relative to the pump pressure?

Memorandum 2

FANNING INDUSTRIES LTD

to M. Grace
Chief Development Engineer

from P. Thirsk
Field Engineering Manager

date 4th April

Large End Bearing Erosion on PS9 Series Diesels

Field maintenance engineers from this department have, in the last week, detected two cases of erosion of large end bearing shells on early production models of the new PS9 series engines in customers' installations. In both cases, the bearing surface showed erosion damage in the vicinity of the opening of the drilling in the connecting rod which conducts oil to the piston crown. The two engines concerned were amongst the first batch of this type sold, and total running hours are higher than any other similar engines. Operators' records for both engines indicate that oil and filter changes have been carried out in accordance with our recommended schedule and that neither engine has run significantly overspeed at any time (<5% in both cases - well within design safety limits). If the erosion problem is due to a design deficiency we may expect an increasing rate of bearing shell damage to be reported, and the work that this necessitates under our warranty will strain our field engineering resources and cause considerable expense to the company. I would appreciate your comments on the problem as soon as possible.

Memorandum 3

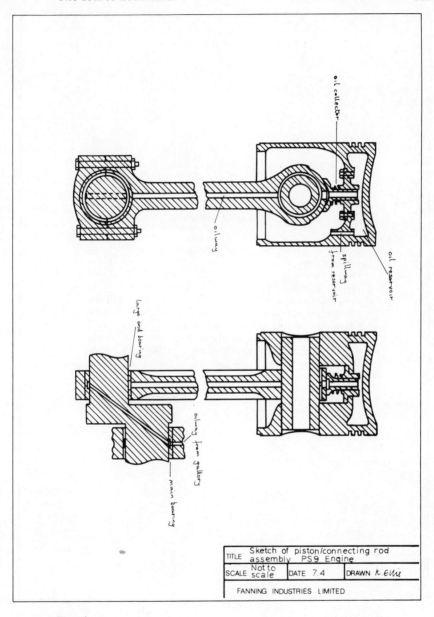

Drawing

PS9 ENGINE DIMENSIONS

Number of cylinders	3
Cylinder bore	0.38m
Piston stroke	0.46m
Connecting rod length	1.15m
Rated operating speed	600 rpm
Lubricating oil inlet temperature	62^0C
Piston cooling oil outlet temperature	81^0C
Main and large end bearing diameter	0.26m
Oil flow through each piston crown	1.566×10^{-3} m^3/s
Oil pressure in crank case gallery	350 KN/m^2
Diameter of connecting rod oilway	0.021m
Diameter of crankshaft oilway	0.030m

Technical details

9.2 Fanning Industries – general lines of donor's solution

The problem explicitly posed to the Analysis Section was the assessment of the inertia forces on the column of oil in the connecting rod drilling. The first stage in making this assessment was the calculation of the accelerations at an arbitrary point in the connecting rod. The force needed to maintain the column of oil in steady motion relative to the connecting rod could then be assessed.

The accelerations were derived through the application of some basic co-ordinate geometry. Using the notation introduced in figure 9.1 it was seen that

$$x = (l - s)\sin(\phi)$$
$$y = r\cos(\theta) + s\cos(\phi).$$

Differentiating these with respect to time produced

$$\dot{x} = (l - s)\cos(\phi)\dot{\phi}$$
$$\dot{y} = -r\sin(\theta)\dot{\theta} - s\sin(\phi)\dot{\phi}.$$

The geometrical constraint

$$l\sin(\phi) = r\sin(\theta)$$

Fig. 9.1. General arrangement of crankshaft and connecting rod.

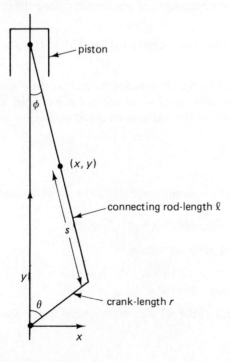

was differentiated and used to eliminate $\dot{\phi}$ thus

$$\dot{\phi} = r\dot{\theta}\cos(\theta)/l\cos(\phi).$$

Assuming the engine turned at a steady speed $\dot{\theta} = \omega$ and introducing non-dimensional parameters $\varepsilon = r/l$ and $S = s/l$ the equations

$$\ddot{x} = -\omega^2(l-s)\sin(\phi) = -r\omega^2(1-S)\sin(\theta) \qquad (9.1a)$$

$$\ddot{y} = -r\omega^2[\cos(\theta) + \varepsilon S\cos(2\theta)/\cos(\phi) + \varepsilon^3 S\sin^2(2\theta)/4\cos^3(\phi)] \qquad (9.1b)$$

were obtained. These equations are relatively complex and intractable in this form. It was noted, however, that the parameter ε has a value of approximately 0.2. If the expressions were expanded as power series in ε terms of $O(\varepsilon^2)$ should be negligible. The Fanning Industries team took this route. This allowed the angle ϕ to be eliminated from the expression for $\ddot{y}$ by using the relations

$$\cos(\phi) = (1 - \varepsilon^2\sin^2(\theta))^{\frac{1}{2}}$$

and hence, using the binomial expansion,

$$1/\cos(\phi) = 1 + 0.5\varepsilon^2\sin^2(\theta) + O(\varepsilon^4)$$
$$1/\cos^3(\phi) = 1 + 1.5\varepsilon^2\sin^2(\theta) + O(\varepsilon^4).$$

As a result, to $O(\varepsilon)$, the results

$$\ddot{x} = -r\omega^2(1-S)\sin(\theta)$$
$$\ddot{y} = -r\omega^2[\cos(\theta) + \varepsilon S\cos(2\theta)]$$

were obtained. The component of acceleration along the connecting rod was then

$$-\ddot{x}\sin(\phi) + \ddot{y}\cos(\phi) = -r\omega^2[\cos(\theta) + \varepsilon[S\cos(2\theta) - (1-S)\sin^2(\theta)]] \qquad (9.2)$$

again to $O(\varepsilon)$.

The pressure differential, P, between the ends of the connecting rod needed to ensure that the column of oil in the rod maintained a state of uniform motion with respect to the rod was obtained by integrating this expression.

$$P = \int_0^l \rho\ddot{s}\,ds$$

$$= -\rho r\omega^2 l \int_0^1 \cos(\theta) + \varepsilon[S(\cos(2\theta) + \sin^2(\theta)) - \sin^2(\theta)]\,dS \qquad (9.3)$$

$$= -\rho r\omega^2 l[\cos(\theta) + 0.5\varepsilon(1 - 3\sin^2(\theta))]$$

The data provided gave the values

$$r = 0.23\,\text{m} \qquad l = 1.15\,\text{m} \qquad \therefore \quad \varepsilon = 0.2$$
$$\omega = 600\,\text{rpm} = 20\pi\,\text{rad s}^{-1}.$$

The specific gravity of the oil was not recorded in the documentation but a reasonable estimate of the specific gravity would be 0.86 giving $\rho = 860 \, \text{kg m}^{-3}$. Hence

$$P = -898 \times 10^3 [\cos(\theta) + 0.1(1 - 3\sin^2(\theta))]. \tag{9.4}$$

The maximum and minimum values of this were found to be 808 kN m^{-2} and -988 kN m^{-2}. It was thus evident that the extremes of pressure which would be required to maintain the oil in constant motion relative to the connecting rod whilst the latter accelerates and decelerates were considerably larger than the oil gallery pressure of 350 kN m^{-2}. The oil will therefore have significant variations of motion relative to the connecting rod.

The analysis thus far had confirmed that the explanation put forward by Mr Grace was, at the least, not impossible. The next stage of the Fanning Industries analysis was to formulate a more complete model of the oil flow velocity in the connecting rod. In this model the acceleration and deceleration of the body of oil relative to the connecting rod was accounted for. The connecting rod itself is an accelerating frame of reference. Using the notation of figure 9.2, the equation of motion for a small element of the oil column could be derived thus

$$\rho A \delta s (\ddot{z} + \ddot{s}) = A[p(s,t) - p(s + \delta s, t) - k \dot{s} \delta s]$$

where $p(s,t)$ is the instantaneous pressure in the oil column, i.e., letting $\delta s \to 0$,

$$\rho(\ddot{z} + \ddot{s}) = -\frac{\partial p}{\partial s} - k\dot{s}.$$

Here $\ddot{z}$ is the longitudinal acceleration of the connecting rod with respect to inertial axes and $\ddot{s}$ is the acceleration of the oil with respect to the connecting rod. The cross-sectional area of the oilway is denoted by A and the term in $k\dot{s}$ is a viscous resistance term with k the resistance coefficient which was unknown at this stage. Since the oilway in the connecting rod has constant cross-section continuity of the oil implied that

$$\frac{\partial v}{\partial s} = 0 \quad \therefore \quad v = v(t)$$

where $v = \dot{s}$. Hence, integrating along the connecting rod, an equation relating

Fig. 9.2. Notation for connecting rod pressure differential calculation.

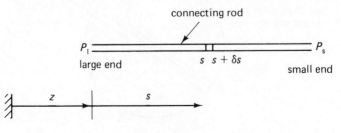

the pressure differential across the ends of the connecting rod and the velocity of oil flow along it was obtained.

$$-\int_0^l \frac{\partial p}{\partial s}\,\mathrm{d}s = \int_0^l \rho(\ddot{z}+\dot{v})+kv\,\mathrm{d}s$$

i.e.

$$P_1 - P_s = l(kv + \rho\dot{v}) + \rho\int_0^l \ddot{z}\,\mathrm{d}s$$

$$= l\,(kv + \rho\dot{v}) - \rho r\omega^2 l[\cos(\theta)+0.5\varepsilon(1-3\sin^2(\theta))],$$

where P_1 and P_s are the pressures at the large end and the small end of the connecting rod respectively. It was assumed that the pressure at the small end would be effectively atmospheric. All other pressures are measured as increments over atmospheric so P_s is the reference pressure (zero).

An effect that is not immediately obvious is the centrifugal pumping action caused by the rotation of the crankshaft. Oil is fed into the oilway in the crankshaft from the main bearings. Thence it is delivered to the large end bearings and so to the oilways in the connecting rods. The rotation of the crankshaft causes the oil pressure in the large end bearing to exceed that in the main bearings. In the notation of figure 9.3

$$P_1 - P_m = \int_{R_2}^{R_1} \rho r\dot{\theta}^2\,\mathrm{d}r$$

$$= 0.5\rho\omega^2(R_1^2 - R_2^2)$$

where P_m is the pressure in the main bearing, which was assumed equal to gallery pressure. R_2 is half the main bearing diameter (i.e. $0.13\,\mathrm{m}$) and R_1 is $R_2 +$ half the stroke (i.e. $0.36\,\mathrm{m}$) so $P_1 = 541\,\mathrm{kN\,m^{-2}}$.

Fig. 9.3. Notation for centrifugal pumping calculation.

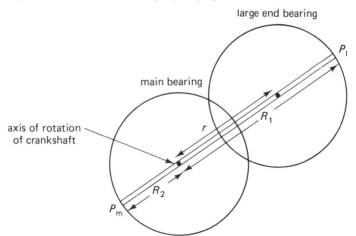

The oil flow velocity relative to the connecting rod was therefore governed by the differential equation

$$\rho l\dot{v} = -klv + P_1 + \rho lr\omega^2[\cos(\theta) + 0.5\varepsilon(1 - 3\sin^2(\theta))]. \tag{9.5}$$

The value of the viscous resistance coefficient k was not known. Its value was inferred from the known oil volume flow rate. Since

$$\dot{v} = \frac{dv}{dt} = \frac{dv}{d\theta}\dot{\theta} = \omega\frac{dv}{d\theta}$$

the differential equation for v was written as

$$\rho l\omega\frac{dv}{d\theta} = -klv + P_1 + \rho lr\omega^2[\cos(\theta) + 0.5\varepsilon(1 - 3\sin^2(\theta))]. \tag{9.6}$$

Integrating over one complete revolution of the fully developed flow in the connecting rod produced

$$0 = -klv_{mean} + P_1 - 0.25\rho lr\omega^2\varepsilon. \tag{9.7}$$

The mean flow velocity in the connecting rod was related to the oil volume flow rate through the relation

$$\text{oil volume flow rate} = \pi a^2 v_{mean}$$

where a is the radius of the drilling in the connecting rod. This gave a value of $k = 95.413 \times 10^3$. A computer dynamic simulation package was used to solve equation (9.5) numerically. It produced the graphs shown in figures 9.4 and

Fig. 9.4. Oil flow velocity, original state and with foot valve.

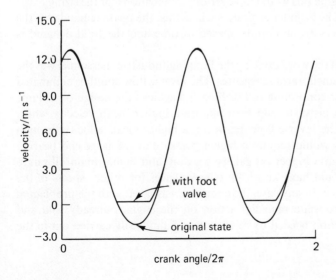

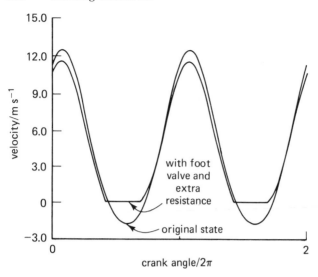

Fig. 9.5. Oil velocity, original and with foot valve plus 11% extra resistance.

9.5. The simulation package was also programmed to compute the oil volume flow rate through the connecting rod as a check on the validity of the argument for estimating the resistance coefficient. The value found was $1.566 \times 10^{-3} \, \text{m}^3 \, \text{s}^{-1}$ thus confirming the validity of the method. The simulation clearly showed that reverse flow does take place during part of the cycle (see figure 9.4), with oil velocities varying between around -1.75 and $12.5 \, \text{m s}^{-1}$. The peak velocity corresponds to a volume flow rate of $4.300 \times 10^{-3} \, \text{m}^3 \, \text{s}^{-1}$, which is nearly three times the mean flow rate. For a single cylinder engine this would have serious implications for the sizing of the oil pump. For multi-cylinder engines, such as this, the peak demands for the individual cylinders are uniformly spaced in time and the total demand is much smoother.

The final part of the work done by the Fanning Industries personnel was the consideration of amelioratory measures. The reverse flow could be eliminated by using a smaller connecting rod drilling. To achieve the design oil volume flow rate to each piston would then require a higher mean velocity which would eliminate the reverse flow. However, a higher mean velocity through narrower oilways would require a higher pressure at the large end bearing which would mean a higher oil gallery pressure and higher main oil pump pressure. This would have knock-on consequences for other aspects of the design of the engine. In any case, the company was faced with the problem of determining appropriate remedial action on the engines already sold and installed. Such action needed to be economical and easily carried out in the field if possible.

The solution chosen was to fit in the base of each connecting rod oilway a small non-return valve (referred to as a foot valve). This valve prevented the reverse flow of oil near the bottom dead centre position. The second trace on figure 9.4 shows the oil flow velocity computed for this configuration. It is seen that the velocities are little changed except in the region of bottom dead centre. The result of this is, however, to increase the mean oil volume flow rate. The simulation showed this to rise to $1.720 \times 10^{-3} \text{ m}^3 \text{ s}^{-1}$. This increase in volume flow rate was not necessarily desirable since it could cause the volumetric capacity of the pump to be exceeded. To bring the volume flow rate back nearer the design figure extra resistance to the flow of oil through the connecting rod oilway was needed. This was provided by inserting a small orifice valve in the exit way from the oil reservoir in the piston crown. This would have the effect of indirectly increasing the resistance to the flow of oil in the connecting rod. Some exploratory simulation runs showed that an increase of 11% in the resistance coefficient k would reduce the oil volume flow rate to $1.558 \times 10^{-3} \text{ m}^3 \text{ s}^{-1}$, near enough to the original figure. Figure 9.5 shows the oil velocity in this new state compared to the original, unmodified design. That did not, of course, directly answer the question of what size of orifice valve was appropriate but gave the design engineers some measure of the effect they were trying to achieve.

9.3 Fanning Industries – commentary on student solutions

One of the interesting aspects of this case study, and one which causes some consternation amongst students, is the data provided. Students are used to being given all the data that is needed for a given problem and only the data needed. In the real world this does not often happen. The list of data provided by the Drawing Office (see memo from Mr James to Mr Fellows) is compiled speculatively so there is no guarantee either that it is complete or that every item is necessary. For the model chosen by the Fanning Industries personnel some of the data is redundant whilst the density of the lubricating and cooling oil, which is needed, is not specified. The author has taught groups of students who refused to proceed further with this problem until he provided an authoritative value for oil density. This is, in fact, unnecessary. Common sense will estimate the value of the specific gravity of the oil between say 0.7 and 1.0. The choice of any value within this range leads to the same initial conclusions. The latter part of the calculations would be modified by a different value for oil density but the conclusions would be qualitatively similar. It is an important lesson for students to learn that parameters used in mathematical models need only be known sufficiently accurately for the purposes in hand. In this case the initial purpose was to compare the order of magnitude of the inertia forces and the oil pump pressure. The conclusion that the inertia forces exceed the pump pressure by a substantial margin is not altered by changing the value of oil density to any value in the likely range.

Another point of interest is the approximation used in the calculation of the accelerations. The expansion of quantities as power series in the parameter ε and the neglect of terms of $O(\varepsilon^2)$ can be avoided. Equation (9.2) is replaced by

$$
\begin{aligned}
-\ddot{x}\sin(\phi)+\ddot{y}\cos(\phi)= & -r\omega^2[\cos(\theta)(1-\varepsilon^2\sin^2(\theta))^{\frac{1}{2}} \\
& +\varepsilon[S\cos(2\theta)-(1-S)\sin^2(\theta)] \\
& +0.25\varepsilon^3 S\sin^2(2\theta)(1-\varepsilon^2\sin^2(\theta))^{-1}]
\end{aligned} \tag{9.2'}
$$

equation (9.3) by

$$
\begin{aligned}
P= & -\rho r\omega^2 l[\cos(\theta)(1-\varepsilon^2\sin^2(\theta))^{\frac{1}{2}} \\
& +0.5\varepsilon(1-3\sin^2(\theta))+0.125\varepsilon^3\sin^2(2\theta)(1-\varepsilon^2\sin^2(\theta))^{-1}]
\end{aligned} \tag{9.3'}
$$

and equation (9.4) by

$$
\begin{aligned}
P= & -898\times10^3[\cos(\theta)(1-0.04\sin^2(\theta))^{\frac{1}{2}} \\
& +0.1(1-3\sin^2(\theta))+0.001\sin^2(2\theta)(1-0.04\sin^2(\theta))^{-1}].
\end{aligned} \tag{9.4'}
$$

The maximum and minimum values of the pressure P are unchanged. The simulation work leading to oil flow velocity profiles is readily repeated with this more complex expression. As would be expected from the magnitudes of the coefficients multiplying the additional terms the difference is undetectable. The approximations made in the donor's analysis were evidently both sensible and valid, allowing the analysis to proceed more easily by eliminating unnecessary complexity from the expressions.

One problem which has frequently arisen with student groups working on this problem concerns the nature of the resistance term to oil flow through the connecting rod oilway. Initially it is quite common for groups to decide to neglect the resistance altogether. This is quite a natural conclusion for them since they are given no data on the resistance coefficient and, as suggested before, students are used to the idea that if data is given it is needed, if it is not given then it isn't needed. They do not immediately see that a resistance coefficient can be derived indirectly and conclude that its absence indicates that it is not necessary. Of course, physical arguments suggest that, in the absence of a dissipative mechanism like resistance, the oil column will, in theory, continue to accelerate indefinitely. Simulations without resistive terms have shown this phenomenon – to the considerable consternation of the student groups involved!

Once it has been decided to include a resistive term it is necessary to decide how resistance varies with velocity. Groups have suggested using square law resistance and various non-integer power laws. These alternatives have the disadvantage that the averaging process used to derive equation (9.7) and so infer a value for the resistance coefficient k no longer produces an equation in v_{mean} but instead one in an averaged power of v_{mean}. This cannot easily be related to the mean volume flow rate of the oil so the argument previously used fails. An appropriate resistance coefficient can be found under these circumstances by repeated simulations using different values of resistance. The appropriate value of k is the one leading to the correct averaged volume flow rate. There does not seem to be any immediately compelling argument for any given resistance law and the author has, in the past, been prepared to accept student suggestions of any power law between about 0.5 and 2. Suggestions outside this range seem less likely and a student group would need to argue strongly and produce supporting evidence for their model to persuade the author that such a choice was sensible and valid.

The simulations of the oil flow used to illustrate this chapter were carried out using a dynamic simulation package. Student groups working on this case study have, over the years, both used simulation packages and written their own programs for the simulation. The problem is a fairly simple one to program. Equation (9.5) is an initial value ordinary differential equation problem easily solved by any standard method. The only slightly problematical aspect of the problem in either case is the initial condition to use. It is the periodic oscillations in the steady state to which the system tends eventually which are of interest so it is convenient to choose an initial condition that results in the system achieving its steady state as rapidly as possible. In fact a little analysis shows that the time constant of the equation is sufficiently short that the initial condition is not very important. The equation (9.5) is of the form

$$\dot{v} = -kv/\rho + f(\theta).$$

The time constant for the transient solutions of this equation is ρ/k which, in this case, has a value of about 0.01 s – in other words 10% of the period. This suggests that the system will have effectively achieved its steady state in well under its first half cycle of operation. This is obviously a very convenient property of the system under study. In the author's experience most of the student groups who get as far as simulating the system start from some arbitrary initial conditions and hope that the system will tend to its steady state sufficiently rapidly. Very few have realised that some prediction of the rate of approach to the steady state can be made. It is obviously worth alerting them to this possibility.

10

Winchester Aircraft Company

10.1 Winchester Aircraft Company – source documents

Aircraft	Administration	Page	1
		Volume	
To T. Fellows	Winchester Aircraft Company	Date 24th March	
From H. Trump	Refs 21CO1/HT/JB	Title Ground Running Bays	

I have attached an extract from the report of the Technical Liaison Meeting actioning the Management Services Department to investigate the delays being experienced in the ground running bays.

I think we can probably construct some sort of model of the problem using queueing theory. It will presumably be necessary to obtain some appropriate data (typical duration of tests on the two types of aircraft, and so on) from the Flight Test Department. Please investigate the modelling problem, determine the data we shall have to obtain and report back to me. It will obviously be particularly useful if we can predict the delays which will result from the anticipated future demand for test bay facilities and what reductions in these delays will be achieved by providing extra bays.

Memorandum 1, page 1

Aircraft	Administration		Page 2
			Volume
To T. Fellows	Winchester Aircraft Company		Date 24th March
From H. Trump	Refs 21CO1/HT/JB	Title Extract from Minutes of Tech'cal Liaison Meeting, 16 March	

A. R. Walters reported that members of the Flight Testing Department were experiencing considerable delays in getting aircraft into ground running bays for engine tests. Engineers had waited for short periods in the past, but recently delays of over 2 hours had been experienced and this was not acceptable. Mr Walters thought that the situation would not improve in the forseeable future, particularly as the Hornet was entering the engine development phase. P. Turner indicated that there would be a small decrease in the demand for testing the Wasp; engine development is now complete and only checkout tests on production aircraft will be needed. Even with this slight decrease in pressure for the test facilities, most heads of departments agreed with Mr Walters' prediction of a higher overall demand for tests to be carried out in the ground running bays. Mr Walters suggested that extra bays should be added to the three currently in use so that the increased demand would not result in further increase in the delays caused to aircraft awaiting engine tests. J. L. Humphries expressed his concern about the cost of providing more ground running bays, semi-buried test facilities such as these being extremely expensive to construct and equip.

Mr Walters felt that it was vital to increase the facilities, but could not provide the meeting with any firm data showing how long engineers were actually having to wait nor how many additional test bays were required to alleviate the situation. Mr Humphries instructed H. J. Trump to investigate this problem and submit a written report in advance of the next meeting.

Memorandum 1, page 2

Aircraft	Technical Office	Page 1
Hornet Wasp		Volume
Prepd By A. Jones Flight Test Dept	Winchester Aircraft Company	Date 3rd April
Title Current Usage of Ground Running Bays	Refs 21CO1	

Tests on two aircraft types are currently being carried out:

HORNET

Six prototypes of this aircraft are currently being flight tested. Aircraft use the ground running bays for checks on engines after repairs, modifications or engine changes. The Hornet project team keep fairly complete records of when their aircraft are in the bays so they may be able to provide you with the data you need.

WASP

The Mk I Wasp entered production 18 months ago and Mk II is currently nearing the end of development and will go into production soon. Engine tests thus fall into two categories: pre-delivery checks on Mk I production aircraft and tests on the Mk II development aircraft. No records of times in and out of the test bays for the Wasp aircraft have been kept as such. The Task Accounting Section will have an idea of the total hours/month spent on testing as we make weekly returns to them for task control and costing purposes. Tests on the Wasp engines are a lot shorter, on average, than those on the Hornet's engines as the Wasp is a simpler aeroplane and the engines are at a later stage of development and so intrinsically more reliable. My estimate of the average test would be about $2\frac{1}{4}$ hours, although I wouldn't like to say how exact that is. None of the work tends to be very urgent, so Wasp tests do tend to cease promptly at 1700 hours, unlike the Hornet tests which often go on until all hours, depending on the urgency of the job.

Memorandum 2

Aircraft Hornet Wasp	Administration	Page 　　　1
		Volume
To 　T. Fellows	Winchester Aircraft Company	Date 　　6th April
From G. Clough Task Accounting Sect	Refs 　　21CO1	Title 　　Ground Engine Tests

　　　The figures we have available are the total test times
booked to each internal accounting task number during each four
week period. We have no information on the duration or dates of
individual tests. We also have the estimates made, for each
task, of the future requirements for testing time. This
information is collected on a routine basis to assist in
estimating future expenditure on each task. We can give no
guidance as to the accuracy of the estimates; if you need further
information you should refer to the appropriate department.

Task 1496 - engine testing of Wasp I production aircraft.
Task 1623 - engine testing for Wasp II development work.
Task 2110 - engine testing of Hornet prototypes.

Period ending	Task 1496	Task 1623	Task 2110
28 January	118 hours	54 hours	174 hours
26 February	124 hours	108 hours	160 hours
26 March	136 hours	112 hours	171 hours
23 April*	120 hours	20 hours	450 hours
21 May*	120 hours	20 hours	450 hours

* estimates of future requirements

Memorandum 3

Aircraft Hornet Prototype	Technical Office	Page 1
		Volume
Prepd By A. Greenfield	Winchester Aircraft Company	Date 5th April
Title Record of Ground Running Bay Usage - 4 weeks ended 27 March	Refs 21CO1	

Ground Running Bay A			Ground Running Bay B			Ground Running Bay C			
Time In	Time Out	Aircraft Number	Time In	Time Out	Aircraft Number	Time In	Time Out	Aircraft Number	Date
0800	1155	R112				1600	1820	R120	1/3
			1005	1930	R117				2/3
			0800	1100	R117				3/3
						1400	1900	R110	4/3
						0800	1430	R110	5/3
1510	1840	R119							6/3
			0800	1200	R111				8/3
			1605	1835	R117				
			2110	2240	R110				
						1145	1625	R120	9/3
0825	1920	R112							10/3
									11/3
						0800	1855	R111	12/3
						1000	1755	R110	13/3
									15/3
									16/3
									17/3
1355	1640	R110	0815	2025	R120	1730	1915	R117	18/3
			0800	1700	R120				
						0800	1055	R119	19/3
1050	1910	R111	1115	1940	R112	1430	1610	R110	20/3
1040	1700	R112							22/3
									23/3
1425	1820	R111	1420	1610	R120	1040	2250	R117	24/3
1235	2040	R120	0900	1055	R112	1230	1440	R119	25/3
			1700	1830	R117				
0800	1230	R120				0805	1540	R111	26/3
									27/3

Memorandum 4

10.2 Winchester Aircraft Company – general lines of donor's solution

This problem, as suggested in the memorandum from Trump to Fellows, has many of the characteristics of a queueing problem. The Management Services Department therefore initially tackled it using queueing theory. Standard models exist for various types of queue but none of those available corresponded exactly to the queue problem facing Winchester Aircraft. Their problem might be described as a multiple-server, single queue with two disparate types of queue member. One of the standard models available is a multiple-server, single queue but the model assumes that all the queue members (or potential queue members) are identical in their properties. It was decided therefore to use this simple model, making an attempt to define some sort of average queue member whose characteristics were a weighted average of the two disparate types of queue member actually present.

The standard model used was that of a queue with c servers. The inter-arrival times of the queue members were taken to be randomly distributed with negative exponential distribution and the service times also negative exponentially distributed. The queue discipline was assumed to be first-in-first-out and the pool of potential queue members assumed infinite. Such a queue is known as an $M/M/c$ queue (see, for instance, Gross and Harris, 1974). Obviously the assumed distributions of inter-arrival and service times were not entirely correct, nor was the assumption of an infinite pool of potential queue members. Nonetheless the theoretical model was taken to be a reasonable guide to what might be expected in practice.

One point of interest was the decision as to which properties of the queue were to be used to evaluate the situation. Obviously the probability of having to queue at all and the mean waiting time were useful. However it is in the nature of human queuers that they complain most vociferously about the occasional long wait even though their average wait may be quite short. This implies that a waiting time probability density function with a long upper tail will give rise to more discontent than another one with the same mean but a shorter upper tail. A statistic of the queue which characterises this property is the probability of having to wait some relatively long time or, alternatively perhaps, the waiting time that is exceeded with some relatively small probability. The waiting time which was exceeded with a 10% probability was chosen as this measure.

The properties of the queue members needed for the model are the rate of arrival of the queue members at the queue, λ, and the service rate for each server, μ. Denoting the number of servers by c, the parameter r is defined as

$$r = \lambda/c\mu.$$

Clearly the parameter r is a measure of the capacity of the system to meet the demand placed upon it. If the queue is not to grow indefinitely, r must be less than unity. Relevant results for the measures of performance desired (see Gross and Harris, *op. cit.*) are

$$W = \frac{(rc)^c P}{c\mu c!(1-r)^2},$$ (10.1)

$$W(0) = 1 - \frac{(rc)^c P}{c!(1-r)},$$ (10.2)

$$W(t) = 1 - \frac{(rc)^c \exp(-c\mu(1-r)t)P}{c!(1-r)}, \quad t > 0$$ (10.3)

where W is the mean waiting time, $W(0)$ is the probability that an arrival at the queue goes immediately to service without waiting, $W(t)$ is the probability of waiting a time t or less in the queue and

$$P = \left[\frac{(rc)^c}{c!(1-r)} + \sum_{n=0}^{c-1} \frac{(rc)^n}{n!} \right]^{-1}.$$ (10.4)

Thus T, the queueing time which is exceeded with probability 0.1, is given by

$$T = \frac{[c\ln(rc) + \ln(P) - \ln(0.1c!(1-r))]}{c\mu(1-r)}.$$ (10.5)

The main problem, in this case, was to estimate appropriate values of λ and μ for the queue members. Since these queue members were not homogeneous it was necessary to create a mythical single queue member type whose properties were defined in terms of properties for the two aircraft types. The definitions

H_h = total test hours for Hornet in period
T_h = mean test duration for Hornet
N_h = number of Hornet tests in period

and similarly for Wasp were used. Then μ was defined by

$$\mu = 1/(\text{mean test time}) = (N_h + N_w)/(H_h + H_w).$$ (10.6)

In order to determine the arrival rate it was necessary to know both the total number of arrivals in some fixed period and the length of that period in working hours. This is complicated for the Hornets by the fact that overtime is worked as necessary on these aircraft. From the Hornet test record sheet it was discovered that 32 tests took place in a total of 173.08 hours. During the four weeks 42.75 hours of overtime were worked (overtime being defined as work after 5 pm). The overtime worked on a given day was defined as the overtime worked by the last test bay to finish, thus neglecting overtime worked in parallel on more than one bay. This gave a figure of 37 hours overtime in 24 days. The mean test day, for Hornets, was thus defined as $9 + 37/24 = 10.54$ hours. For Wasp tests the mean test day was 9 hours, the normal working day.

Thus a definition of a mean test day was taken to be

$$\text{mean test day} = (10.54H_h + 9H_w)/(H_h + H_w),$$

the weighted average of the two. From this result were derived

$$\text{mean arrival interval} = (24 \times \text{test day})/(N_h + N_w),$$

and

$$\lambda = 1/(\text{mean arrival interval})$$
$$= (H_h + H_w)(N_h + N_w)/(24(10.54H_h + 9H_w)). \tag{10.7}$$

The memorandum from the task accounting section showed that there had been a significant increase in the test hours on Task 1623 approximately two months before the date of the memo. At approximately the date of the memo the test demands under task 1623 fell but those under task 2110 rose considerably. It was decided, therefore, to predict the properties of the queue for three periods, up to 28.1.76, for 31.1.76 to 26.3.76 and from 29.3.76 onwards.

For the first period $H_h = 174$ hours and $H_w = 172$ hours. N_h and N_w were estimated by dividing H_h by T_h and H_w by T_w respectively. The mean test durations were $T_h = 5.41$ hours (estimates from the 32 tests in 173 hours derived from the ground running bay usage record) and $T_w = 2.25$ hours (the best information available for this was the memo from A. Jones in the flight test department) giving values of $N_h = 32.2$ and $N_w = 76.4$. From these figures were derived the values $\lambda = 0.463$ arrivals per hour and $\mu = 0.314$ services per hour and hence $r = 0.492$. Finally the statistics $W(0) = 0.77$, $W = 29$ minutes and $T = 1.72$ hours were obtained.

For the second period similar computations resulted in $r = 0.585$, $W(0) = 0.66$, $W = 48$ minutes and $T = 2.87$ hours and for the final period $r = 0.805$, $W(0) = 0.34$, $W = 4.56$ hours and $T = 13.07$ hours.

The queueing theory model made predictions that could be summarised as follows. During the first period the probability of finding an empty bay when one was needed was a little better than 75%, the mean wait experienced was around half an hour and on only 10% of occasions did engineers have to wait more than an hour and three quarters for a test bay. However, in the last two months, the chances of finding an empty bay have fallen to about 66%, the average wait has risen to three quarters of an hour and the wait exceeded on 10% of occasions has risen to nearly three hours. The implications of the memoranda are that this situation is giving rise to a level of dissatisfaction amongst the test engineers. The mean wait has only risen by 15 minutes, which seems pretty harmless, but the more telling change is the wait exceeded on 10% of occasions. This has nearly doubled to around three hours. Predictions for the future made with the same model suggest that the chance of finding an empty bay is likely to fall to about one in three and the mean wait will rise to around four and a half hours. This is obviously not acceptable from either an efficiency or morale point of view.

The model allowed the investigation of the changes that would result from the provision of extra engine ground running bays. Using the above expressions with $c = 4$ gave $W(0) = 0.71$, $W = 45$ minutes and $T = 2.75$ hours, that is somewhat better than the situation that prevailed during the last two months but not as favourable as the situation prevailing three months ago. Increasing the number of test bays to five produced $W(0) = 0.88$, $W = 11$ minutes and $T = 14$ minutes, considerably better even than the situation prevailing three months ago. The indication of this work was that, to meet the demand for the next few months and beyond, one extra ground running bay is almost essential and a second additional bay may also be justified.

This case study serves very well to illustrate the difficulties that often arise when analysing real problems. It is often the case that there exists a standard mathematical model of a similar situation but the model cannot easily be extended to take account of the features of the real problem which do not exactly match the model. In such cases it may be possible, by making suitable approximations of reality, to ignore the features making the real problem different. Obviously such a process is somewhat unsatisfactory from an academic point of view but is often the only way to proceed in solving real industrial or commercial problems.

10.3 Winchester Aircraft Company – commentary on student solutions

The main alternative route for the solution of this problem is the use of computer simulation. Discrete event simulation systems, such as SIMULA, or custom built simulation programs could be used. Simulation has the advantage that the main complicating feature of this problem, the existence of two disparate queue member types, can be included in the simulation. The use of simulation, however, raises the issue of reliability of the results of simulation, that is how many weeks or months of operation must be simulated using each set of parameters in order to obtain reliable results. In the author's experience those students who have chosen to take the simulation route do not fully anticipate the problems that simulation and the interpretation of its results brings. Rather simulation is seen as an easy way of dealing with an analytically intractable problem. This, of course, is not entirely the case as they find to their cost.

The other feature worth noting here was originated by a group of students who were worried about the accuracy of the estimate of the Wasp mean service time (2.25 hours). In view of their concern they were asked to conduct a sensitivity analysis of their results for variations of the Wasp service time. Sensitivity analysis is a very powerful tool in the armoury of the industrial or commercial mathematician. If the value of a parameter needed in a mathematical model is suspect then action may be necessary to confirm the value. It is wise though, first of all, to check the sensitivity of the results or conclusions of the analysis to changes in the value of that parameter. If the value given for a parameter is suspect but a variation of that value by 50% only changes the value of the result by 5% say, then the lack of accuracy of the parameter value is unlikely to be of any great concern. On the other hand, if the result varies by a larger percentage than the variation of the parameter an accurate knowledge of the parameter value is probably important.

The sensitivity analysis performed in this case was to recompute all the results quoted above using Wasp mean service times of 2 hours and 2.5 hours. This is a variation of about 11% in the parameter. The largest variation of any result was a variation of 9% and, in most cases, the variation of the result was of the order of 5%–6%. The probability of not waiting at all, $W(0)$, was unchanged in all cases. Overall, therefore, the results did not appear to be particularly sensitive to this input parameter.

An alternative way of carrying out a sensitivity analysis is to calculate the differentials of the quantities involved with respect to the parameters to be varied. This procedure is only viable if the analytic expressions involved are not too complex. It gives rather more information but, because of the additional complexity, is rather less likely to be used in the real world than the alternative method outlined above. In this case, combining equations 10.6 and 10.7 shows that the quantity r is independent of the Wasp mean test duration, T_w. As a result equation (10.4) shows that P is independent of T_w and then equation (10.2) shows that $W(0)$ is also independent of T_w. This explains the insensitivity of $W(0)$ to T_w noted above. Equations (10.1) and (10.5) yield the results

$$\frac{dW}{W} = -\frac{d\mu}{\mu} \quad \text{and} \quad \frac{dT}{T} = -\frac{d\mu}{\mu}.$$

From equation (10.6) and the subsidiary result $N_w = H_w/T_w$, an expression for $d\mu/\mu$ can be obtained

$$\frac{d\mu}{\mu} = -\frac{N_w}{(N_h + N_w)} \frac{dT_w}{T_w}.$$

Combining these leads to

$$\frac{dW}{W} = \frac{dT}{T} = \frac{N_w}{(N_h + N_w)} \frac{dT_w}{T_w}.$$

The form of the expression demonstrates that the fractional change in W and T will always be less than the fractional change in T_w.

11

Natural Gas

11.1 Natural Gas – source documents

NATURAL

GAS

LIMITED

MEMORANDUM

FROM	P. Thomas Advance Projects Division	TO	T. Fellows Mathematical Section
TELEPHONE	0593 812 613	DATE	18th May
SUBJECT	UNDERGROUND STORAGE OF GAS	REF	PT/DGH/5819

It has been proposed by Dr. Kitcher of the Engineering Laboratory that underground salt strata could be exploited for the storage, under high pressure, of natural gas. The scheme involves drilling bore holes of approximately 6-inch diameter down to the salt strata. A cavity in the strata would then be formed by "leaching" of the salt, that is, pumping water down the borehole through the inner duct of a concentric pipe, so dissolving the salt. The brine formed would be withdrawn up the borehole in the outer duct of the pipe. It is anticipated that a roughly spherical cavity would be formed in the salt.

Subsequent to formation, the cavity will, to a certain extent, close up due to creep, under pressure, of the surrounding strata. The extent of the closure will depend on the pressure of the gas stored in the cavity. It is proposed to determine the extent of the likely closure by drilling a pilot cavity of approximately 60m³ volume for *in situ* tests. The purpose of these tests will be to measure cavity volumetric closure against cavity internal pressure. Ideally, the cavity will be held at constant pressure whilst the variation of cavity volume with time is observed. A series of tests at different pressures will yield information on the creep behaviour of the strata surrounding the cavity which will be used in the design of the full-scale storage cavities.

The intention is to leach out the cavity, then maintain a constant pressure inside it by controlling the height of the brine in the borehole. The change in cavity volume will be calculated from the volume of the brine added or extracted to maintain the head. However, the volume of the brine extracted will not necessarily be a direct measure of the volumetric change of the cavity with time due to creep under constant pressure. A number of complicating factors are anticipated, the most significant of which is thought to be the change of brine volume due to change of temperature. The salt strata, when undisturbed, are at a temperature of approximately 57°C, whereas the water used to leach out the cavity will be at a temperature of between about 3°C and 12°C, depending on the time of year. The water entering the cavity will thus be significantly cooler than the salt strata surrounding the

Memorandum, page 1

FROM P. Thomas   TO   T. Fellows

- 2 -

cavity. When leaching of the pilot cavity is stopped, the brine
in the cavity will warm up, which will increase the apparent
cavity closure. The magnitude of this effect and the time scale
over which the warming up will take place must be estimated in
order to be able to interpret the data obtained from the pilot
cavity correctly. Two problems have been identified so far:

1. What will be the temperature of the brine in the cavity
when leaching is stopped? The salt dissolved in the water will
initially have been at $57^\circ C$, and the water at around $7.5^\circ C$ (mean
of the two estimated extreme temperatures). The heat of solution
of the salt must also be considered. As the salt dissolves it
absorbs heat and this process will reduce the temperature of the
brine in the cavity. The brine temperature may thus be expected
to be nearer that of the water than that of the salt. Can you
make an accurate estimate of the brine temperature at the end of
leaching?

2. How long will the remaining brine in the cavity take to
reach the temperature of the surrounding salt strata when leaching
is stopped? When leaching ceases, the cavity will be filled with
brine at the temperature indicated by the solution to the above
problem. Over some period the brine will heat up to the temperature of the salt strata surrounding the cavity. It seems likely
that convection within the brine will keep the mass of brine at a
roughly spatially uniform temperature during this process, though
that temperature will vary with time of course. The leaching
process is expected to take between 10 and 20 hours to form a $60m^3$
cavity. This rate of salt removal is expected to be sufficient to
maintain the retreating salt/brine interface at roughly the strata
temperature. (Confirmation of this expectation from theoretical
considerations would be valuable, however). When the leaching
ceases and the salt/brine interface becomes stationary, heat will
flow from the salt to the brine and a temperature gradient in the
salt strata will be set up. This will decay as the brine warms up
to the strata temperature. Can any estimate be made of the time
taken for the brine in the cavity to warm up to the temperature of
the salt strata?

We would appreciate your comments on these two problems as soon
as possible.

Memorandum, page 2

11.2 Natural Gas – general lines of donor's solution

The engineers at Natural Gas Ltd were exploring a novel way of storing natural gas in the interval between its extraction from the offshore gas fields and its consumption by the company's customers. The suggestion was that gas could be stored in artificial underground caves created in rock salt strata below the sea bed. The storage caves were to be made by drilling boreholes down to the salt strata and then pumping water (sea water possibly) down the shafts. The water would dissolve the salt around the bottom of the borehole (a process known as leaching) until the brine became saturated and no more salt would dissolve in it. As long as the saturated brine was continuously pumped out of the borehole and replaced by fresh water, leaching would continue and the cavity would grow. When the cavity reached the desired size the remaining brine would be extracted and gas stored in the cavity under high pressure.

One difficulty anticipated in the implementation of the scheme was the effect of the pressure of the surrounding strata on the rock salt deposits. This pressure was expected to cause the cavity, once leaching was stopped, to close up slowly, reducing the available storage space. The exact degree of closure was expected to depend on the pressure at which the stored gas was maintained. In order to quantify this effect it was suggested that some experimental cavities should be made and the closure of these under various internal pressures could then be measured. A simple way of creating the desired internal pressure would be to leave the cavity full of brine and keep a constant level of brine in the borehole. The higher the level in the borehole the higher the internal pressure in the cavity. Closure of the cavity would be manifest as a rise in the level of the brine in the borehole. In order to keep the internal pressure constant brine would have to be extracted from the borehole. The volume of brine so extracted would be a measure of the closure of the cavity.

This relatively simple experimental scheme was made more complicated by the likely effects of heat transfer. It was known that, at the depth of the salt strata, the ambient temperature is 57 °C. The water used for the leaching would be at a temperature of roughly 7.5 °C. Therefore when leaching was stopped the brine in the cavity would heat up causing it to expand and the level in the borehole to rise. This rise in level due to the increase in volume of the brine needed to be separated from the rise in level due to the contraction of the bore-hole. The first question which arose was how long would it take for the brine in the borehole to reach temperature equilibrium with the surrounding salt strata? If this period were to be sufficiently short (say a few hours or even a few tens of hours) it would be unlikely to cause much problem, since the cavity closure due to creep of the salt strata was expected to take place on a time scale of days and weeks. If, on the other hand, the time scale for equilibration of the brine and salt strata were comparable to the creep time scale the problem would be much more complex. The company's engineers approached the problem on a broad front. Being a large organisation with research departments in different geographical locations they were able to commission different groups to look at various potential solution methods.

The first investigation was in the nature of an engineering feasibility study. The model assumed that the cavity formed would be spherical, that the brine in the cavity would be well mixed (by convection and other effects) and so of uniform properties, and that the temperature gradients set up in the salt strata surrounding the cavity would have spherical symmetry. Physically the expectation on which the model was based was that, during leaching, water at some temperature between about 3 °C and 12 °C would be pumped down the central core of a concentric pipe in the borehole and saturated brine returned up the outer part of the pipe. In the cavity the water would dissolve salt from the surrounding strata. The change of state of the salt from crystalline rock salt (halite) to dissolved salt in the brine requires the supply of heat of solution. As a result the saturated brine would be cooled. At the end of the leaching stage, when the cavity had reached its required size, pumping would be halted. At this point the cavity would be filled with saturated brine at some temperature and the close surrounding salt strata would have some temperature deficit relative to the ambient level. This deficit would reduce, in an approximately exponential manner, to nothing at large distances from the cavity as indicated in figure 11.1. Heat would then flow from the large mass of salt into the region

Fig. 11.1. Temperature profile at end of leaching.

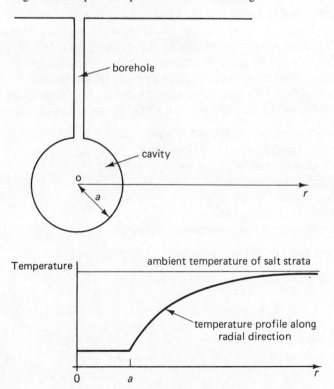

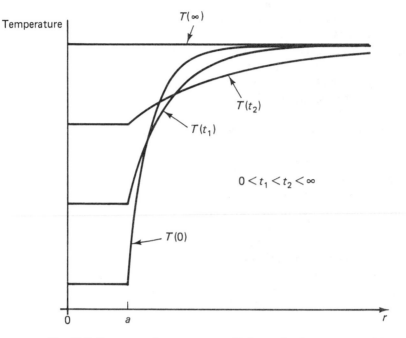

Fig. 11.2. Recovery of temperature of brine and salt strata to ambient.

close to the cavity and thence into the brine in the cavity. This would continue until the whole system finally attained the ambient temperature, as indicated in figure 11.2. The mass (and therefore the heat capacity) of the whole stratum is, of course, sufficiently great that the final ambient temperature would be insignificantly different from the starting ambient temperature.

The first stage of this study, then, was to estimate the temperature of the brine at the end of the leaching process. Defining quantities as follows:

Initial temperature of input water	T_s
Final temperature of saturated brine	T_c
Initial temperature of salt strata	T_b
Specific heat of water	C_w
Specific heat of salt	C_s
Heat of solution of salt	H
Composition of saturated brine by weight	$S\%$ salt,

a simple heat balance of input and output from the cavity gave

$$(100-S)C_w(T_s-T_c)+SC_s(T_b-T_c)=SH.$$

Taking

$$T_s = 7.5\,°C$$
$$C_w = 4.186\,\mathrm{J\,g^{-1}\,°C^{-1}}$$
$$H = 66.40\,\mathrm{J\,g^{-1}}$$

$$T_b = 57\,°C$$
$$C_s = 0.88\,\mathrm{J\,g^{-1}\,°C^{-1}}$$
$$S = 26.4\%$$

yields $T_c = 5.7\,°C$. The solubility of salt in water is strictly a function of the temperature of the resulting brine. The variation is small however, 26.3% at $0\,°C$ rising to 27.2% at $60\,°C$. The value of 26.4% used in the calculation is that pertaining at $10\,°C$. T_s will, of course, show seasonal variations. Taking the suggested maximum and minimum of $12\,°C$ and $3\,°C$ yielded variation of T_c between $9.9\,°C$ and $1.5\,°C$.

The next question to be answered was whether the change of temperature of the brine in the cavity from its value at the end of the leaching process to its final value, the temperature of the surrounding salt strata, would cause noticeable expansion. Data for the density of brine at $25\,°C$ and $40\,°C$ were used to estimate the coefficient of cubical expansion of brine. This was found to be roughly $0.47 \times 10^{-3}\,°C^{-1}$. In heating up from $5.7\,°C$ to $57\,°C$ therefore the brine would expand by 2.4%. For a $60\,\mathrm{m}^3$ cavity this is $1.45\,\mathrm{m}^3$. In a borehole with a diameter of 6 inches this would cause a rise of around $80\,\mathrm{m}$. In any event 2.4% was thought likely to be significant with respect to the likely closure of the cavity due to creep of the salt strata.

The logical next stage of this preliminary study was to estimate the rate of the heating of the brine. It was argued that, at the end of the leaching process, there would be a restricted volume of salt surrounding the cavity which would show a significant temperature deficit. At some radius the temperature would have become negligibly different from that ambient in the salt strata. The changes in the temperature distribution are governed by the spherically symmetric diffusion equation

$$\frac{\partial T}{\partial t} = \frac{\kappa}{\rho C}\left[\frac{\partial^2 T}{\partial r^2} + \frac{2}{r}\frac{\partial T}{\partial r}\right].$$

where κ, ρ and C are the thermal conductivity, density and specific heat of the salt. If the rate of heating were small the temperature distribution in the salt surrounding the cavity would be negligibly different from that predicted by the steady state diffusion equation

$$\frac{\kappa}{\rho C}\left[\frac{\partial^2 T}{\partial r^2} + \frac{2}{r}\frac{\partial T}{\partial r}\right] = 0$$

whose general solution is

$$T = A/r + B. \tag{11.1}$$

It was assumed that the leaching process left a cavity of radius a filled with brine and the salt temperature unchanged at a radius b. The temperature of the brine, which would now be a function of time, was taken to be T_a. The temperature distribution in the salt satisfying these boundary conditions is

$$T = \frac{aT_a(b-r) + bT_b(r-a)}{r(b-a)}.$$

The heat flux across the sphere $r = a$ is given by

$$Q = -4\pi r^2 \kappa \frac{dT}{dr}$$
$$= -4\pi\kappa\frac{(T_a - T_b)ab}{(b-a)}. \qquad (11.2)$$

The rate of temperature rise of the brine in the cavity is then given by Q/J where J is the heat capacity of the brine filled cavity. J is given by

$$J = 4/3\pi a^3\rho_b C_b \qquad (11.3)$$

where

ρ_b = density of brine
C_b = specific heat of brine.

Taking values

$\rho_b = 1200\,\mathrm{kg\,m^{-3}}$ $C_b = 3.31\,\mathrm{J\,g^{-1}\,^\circ C^{-1}}$
$\kappa = 6.3\,\mathrm{W\,m^{-1}\,^\circ C^{-1}}$ $a = 2.43\,\mathrm{m}$ (60 m^3 cavity)

then gave

$$Q/J = 0.0029\frac{(T_b - T_a)b}{(b-2.43)}\,^\circ\mathrm{C\,hr^{-1}}.$$

(The value used for C_b was obtained by weighted averaging the values C_w and C_s appropriately). The radius b was still, of course, to be determined or estimated. This could not be done easily. Instead, order of magnitude estimates of the rate of temperature rise of the brine in the cavity were obtained for a range of values of b. Immediately after leaching ceases, when the rate of temperature rise is greatest, the estimate varied from $78 \times 10^{-6}\,^\circ\mathrm{C\,s^{-1}}$ (0.28 $^\circ$C per hour) when b is assumed 5 m to $44 \times 10^{-6}\,^\circ\mathrm{C\,s^{-1}}$ (0.16 $^\circ$C per hour) when b is assumed 20 m. The argument invoked above, that the rate of heating is small and that quasi-steady state temperature distributions would therefore provide sensible estimates of the actual conditions, was given considerable credibility by these figures.

Equations (11.2) and (11.3) were used to form a differential equation for the time variation of the temperature of the brine in the cavity thus

$$\frac{dT_a}{dt} = \frac{Q}{J} = -\frac{3\kappa b(T_a - T_b)}{(b-a)a^2\rho_b C_b}, \qquad T_a(0) = T_c$$

with solution

$$T_a(t) = T_b + (T_c - T_b)\exp\{-3\kappa bt/[(b-a)a^2\rho_b C_b]\}.$$

From this expression an estimate of the time scale of temperature equilibration was made. Obviously this estimate also depended on the value assumed for the radius b. For b assumed to be 5 m the temperature differential $T_a - T_b$ falls to half its initial value in 4.3 days and to 1 °C in 24.6 days. For *b* assumed to be 20 m $T_a - T_b$ falls to half its initial value in 33.7 days and to 1 °C in 168.3 days.

This initial model obviously indicated that it was extremely unlikely that the temperature of the brine left in the cavity at the end of leaching would equilibrate with its surroundings sufficiently rapidly for variations in the level of the brine in the bore-hole due to temperature change to be clearly separated from those due to cavity closure.

A subsequent refinement of this initial model was undertaken by another mathematician/engineer on Natural Gas' staff. The refinement affected the second phase of the process, the recovery of the temperature of the brine in the cavity to the salt strata ambient temperature. A more accurate analytic model of this phase of the process was obtained. The improved model retained the assumption of spherical symmetry. It was argued that, at the end of the leaching phase, the temperature distribution in the salt around the spherical cavity would lie between two extreme cases. These are shown in figure 11.3. The first extreme would apply if the leaching process were very fast (strictly speaking infinitely fast). In this case the finite thermal conductivity of the salt would mean that the strata surrounding the cavity were, at the end of leaching, uniformly at the ambient temperature of the more distant strata – there would have been insufficient time during the leaching process for the salt to give up any of its heat. The other extreme possible would apply if the leaching process

Fig. 11.3. Alternative extreme temperature conditions at end of leaching phase.

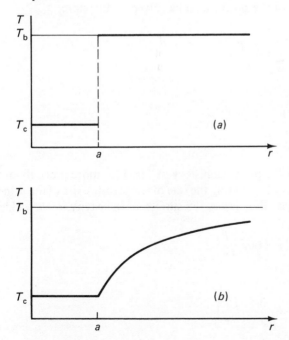

were very slow. In this case the salt strata would be able to reach equilibrium with the brine in the cavity and, at the end of leaching, the salt strata have a temperature distribution which is a solution of the steady state spherically symmetric diffusion equation with a spherical heat sink maintained at constant temperature embedded at the origin. The solution to this, as was noted above (equation (11.1)), is

$$T_0(r) = T_b + (T_c - T_b)a/r$$

where $T_0(r)$ denotes the temperature of the surrounding strata at the end of leaching. Subsequently the temperature in the salt strata, $T(r,t)$, is governed by the time dependent spherically symmetric diffusion equation

$$\frac{\partial T}{\partial t} = \frac{\kappa}{\rho_s C_s} \frac{1}{r^2} \frac{\partial}{\partial r}\left(r^2 \frac{\partial T}{\partial r}\right) \tag{11.4}$$

with appropriate initial condition, κ, ρ_s and C_s denoting the thermal conductivity, density and specific heat of the salt respectively.

The actual initial condition is unknown but, intuitively, it is expected that no achievable initial condition could result in a return to ambient temperature faster than the condition depicted in figure 11.3(a) and similarly no condition could cause return more slowly than that depicted in 11.3(b). The solution of equation (11.4) with both sets of boundary conditions should therefore provide rigorous bounds on the possible temperature distribution histories and particularly on the time taken for the brine to reach the ambient temperature of the surrounding salt strata.

The full statement of the problem to be solved is, therefore,

$$\frac{\partial T}{\partial t} = \frac{\kappa}{\rho_s C_s} \frac{1}{r^2} \frac{\partial}{\partial r}\left(r^2 \frac{\partial T}{\partial r}\right)$$

with

$$
\begin{array}{ll}
T(r,0) = T_0(r) & r > a \\
T(a,t) = T_a(t) & \\
T(r,t) \to T_b \text{ as} & r \to \infty.
\end{array}
$$

But $T_a(t)$ depends on the previous history of $T(r,t)$ (or more precisely of the gradient of $T(r,t)$ at $r = a$). Equating the rate of temperature rise of the brine in the cavity with the heat flux across the surface of the cavity, $r = a$, yields

$$\tfrac{4}{3}\pi a^3 \rho_b C_b \frac{\mathrm{d}T_a}{\mathrm{d}t} = \kappa 4\pi a^2 \frac{\partial T}{\partial r}\bigg|_{r=a}$$

i.e.

$$\frac{\mathrm{d}T_a}{\mathrm{d}t} = \frac{3\kappa}{\rho_b C_b a} \frac{\partial T}{\partial r}\bigg|_{r=a}.$$

The problem was then non-dimensionalised in the usual way. Taking

$$\sigma = \frac{r}{a} \qquad \tau = \frac{\kappa t}{\rho_s C_s a^2} \qquad \Theta = \frac{T - T_b}{T_0(a) - T_b}$$

results in

$$\frac{\partial \Theta}{\partial \tau} = \frac{1}{\sigma^2} \frac{\partial}{\partial \sigma} \left(\sigma^2 \frac{\partial \Theta}{\partial \sigma} \right) \tag{11.5}$$

with

$$\Theta(\sigma,0) = (T_0(a\sigma) - T_b)/(T_0(a) - T_b) \tag{11.5a}$$
$$\Theta(1,\tau) = (T_a(\rho_s C_s a^2 \tau / \kappa) - T_b)/(T_0(a) - T_b) \tag{11.5b}$$
$$\Theta(\sigma,\tau) \to 0 \text{ as } \sigma \to \infty \tag{11.5c}$$

and, since

$$\frac{dT_a}{dt} = \frac{\partial T}{\partial t} \Big|_{r=a},$$

$$\frac{\partial \Theta}{\partial \sigma}(1,\tau) = \alpha \frac{\partial \Theta}{\partial \tau}(1,\tau) \quad \text{with } \alpha = \frac{\rho_b C_b}{3\rho_s C_s}.$$

The two initial conditions, non-dimensionalised appropriately, are

$$\Theta(\sigma,0) = 0 \qquad \sigma > 1 \qquad \text{(figure 11.3a)}$$
$$\Theta(\sigma,0) = 1/\sigma \qquad \sigma > 1 \qquad \text{(figure 11.3b)}$$

so the boundary conditions, equations 11.5a, 11.5b and 11.5c, are finally reduced to

$$\Theta(\sigma,0) = 0 \text{ or } 1/\sigma \text{ as appropriate}$$

$$\frac{\partial \Theta}{\partial \sigma}(1,\tau) = \alpha \frac{\partial \Theta}{\partial \tau}(1,\tau)$$

$$\Theta(\sigma,\tau) \to 0 \text{ as } \sigma \to \infty.$$

This problem was solved using the Laplace transform technique. Defining

$$V(\sigma,s) = \int_0^\infty \Theta(\sigma,\tau) \exp(-s\tau) d\tau$$

the solutions to the two problems were found to be

$$V(\sigma,s) = \frac{i}{2p_2\sigma} \exp(-(\sigma-1)\sqrt{s}) \left[\frac{1}{\sqrt{s+p}} - \frac{1}{\sqrt{s+p^*}} \right]$$

and

$$V(\sigma,s) = \frac{1}{s\sigma} - \frac{i}{2p_2\alpha\sigma} \exp(-(\sigma-1)\sqrt{s}) \left[\frac{1}{s(\sqrt{s+p})} - \frac{1}{s(\sqrt{s+p^*})} \right]$$

where p and $p*$ are defined by

$$p_1 = \frac{1}{2\alpha}, \quad p_2 = \sqrt{\left[\frac{1}{\alpha} - \frac{1}{4\alpha^2}\right]}, \quad p = p_1 + ip_2.$$

These expressions can be inverted to give $\Theta(\sigma,\tau)$ using standard results for Laplace transforms (for instance 29.3.88 and 29.3.89 of Abramowitz and Stegun, 1965). The result actually needed is the values of $\Theta(1,\tau)$. This was found to be

$$\Theta(1,\tau) = \frac{-i}{2p_2}[p\exp(p^2\tau)erfc(p\sqrt{\tau}) - p*\exp(p*^2\tau)erfc(p*\sqrt{\tau})]$$

and

$$\Theta(1,\tau) = \frac{i}{2p_2\alpha}\left[\frac{1}{p}\exp(p^2\tau)erfc(p\sqrt{\tau}) - \frac{1}{p*}\exp(p*^2\tau)erfc(p*\sqrt{\tau})\right].$$

Taking the value $\rho_s = 2170\,\text{kg m}^{-3}$ and using the other values previously given the value of α was found to be 0.693 and hence p was $0.721 + 0.960i$. The two curves for $\Theta(1,\tau)$ are plotted in figure 11.4. From the figure it can be seen that the temperature differential $T_a - T_b$ between the brine in the cavity and the ambient temperature of the salt strata is predicted to fall to 50% of its initial value in something between 3 and 25 days and to 2% (approximately 1 °C) in a period greater than 100 days. This confirmed the initial model's prediction that the time scale of temperature equilibrium would not be short compared with the time scale of creep closure of the cavity.

The third investigation undertaken by the Natural Gas engineers used a numerical approach to the problem. The results of the previous model showed that the time scale of temperature equilibration would be rather sensitive to the temperature distribution in the salt strata surrounding the cavity at the end of the leaching stage. Consequently a simulation of the leaching process was undertaken in an attempt to gain more insight into this factor. The basic assumptions of this model were again that the problem exhibited radial symmetry so that the heat flows would be determined by solutions of the one dimensional radial diffusion equation, (11.4). In this case the boundary conditions were taken to be that the temperature tended to the ambient temperature of the salt strata, T_b, at large distances from the cavity, that the initial temperature of the salt strata was also T_b and that the temperature at the cavity/salt interface was the exit temperature of the brine from the cavity during leaching, T_c. The position of the cavity/salt interface was, however, a known function of time. Thus, in non-dimensional form, the problem was expressed as

$$\frac{\partial\Theta}{\partial\tau} = \frac{1}{\sigma^2}\frac{\partial}{\partial\sigma}\left(\sigma^2\frac{\partial\Theta}{\partial\sigma}\right) \tag{11.6}$$

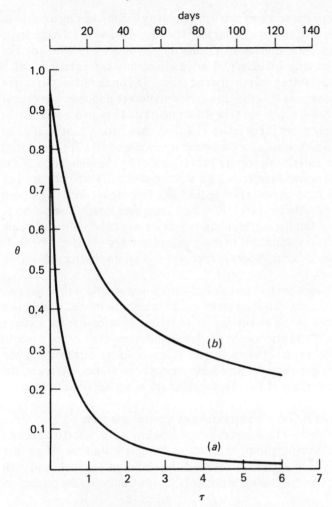

Fig. 11.4. Cavity temperature as a function of time for the two extreme initial conditions of figure 11.3.

with

$$\Theta(\sigma,0)=0, \qquad\qquad \sigma \geqslant \gamma(\tau) \qquad\qquad (11.6a)$$

$$\Theta(\gamma(\tau),\tau) = -1 \qquad\qquad\qquad (11.6b)$$

$$\Theta(\sigma,\tau) \rightarrow \qquad\qquad 0 \text{ as } \sigma \rightarrow \infty \qquad\qquad (11.6c)$$

where $\gamma(\tau)$ is the non-dimensionalised position of the interface.

The function $\gamma(\tau)$ was chosen by assuming that the rate of leaching was constant (in terms of volume of salt removed per unit time) and that the cavity would reach its final size in 10 hours. It was not possible, of course, to use an initial cavity of zero size so a very small initial cavity (0.5 m^3, having a radius of about 20% of the final size) was assumed. The equations were solved using a Crank–Nicholson finite difference scheme appropriately modified to take account of the moving boundary. The result of this computation was a prediction of temperature profile at the end of leaching similar to that shown in figure 11.5. This suggests that the temperature deficit region does not extend very far into the salt strata so that the subsequent equilibrium process should follow more closely curve (a) of figure 11.4 than curve (b). A continuation of the numerical solution using now the boundary conditions (11.5b) and (11.5c) together with the initial temperature profile yielded by the simulation of the leaching process showed that, for $\tau > 0.5$ at least, the cavity temperature was indistinguishable from curve (a) of figure 11.4. This model therefore added further evidence that the time scale of temperature equilibration between the brine in the cavity and the surrounding salt strata would be sufficiently long that changes in brine volume due to temperature changes would not be likely to be easily separable from changes in apparent volume due to creep closure of the cavity.

These investigations were, in fact, only the preliminary actions in a long and complex investigation which went on to use finite element models to simulate the creep behaviour of the salt strata for the 60 m^3 pilot cavity, for a larger experimental cavity and also for the projected full size cavities. Since it would not be realistic to expect undergraduate students to go further than the preliminary investigation in the time scale appropriate to these exercises, the results of the later stages of the investigation are of no importance here.

11.3 Natural Gas – commentary on student solutions

This problem, in its original form, obviously gave rise to a long and complex series of investigations. In order to bring it within the scope of a student case study in this series the source documentation was written in such a way as to direct attention to two aspects of the problem; firstly the prediction

Fig. 11.5. Temperature profile at end of leaching predicted by numerical solution (to same scale as figure 11.3).

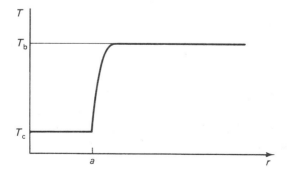

of the brine exit temperature during the leaching process and secondly the development of the temperature profile in the salt strata subsequent to the leaching.

The first obstacle students face is the collection of the physical data needed. This has been used as part of the exercise for the students. Real life problems often exhibit this characteristic and some experience in defining the data needed and using the resources of a university (its libraries and staff) to collect them is of considerable value. In the author's experience two books, Kaye and Laby (1986) and Weast (1985) are invaluable in this context. A related problem is that of units. Where data are collected from several sources (and sometimes even from different sections within a single source) it is almost inevitable that there will be some conflict between the units in which different data are expressed. The resolution of such conflict and the expression of all data in commensurable form is a very valuable exercise and one which students often find surprisingly taxing! The main variations in students' approaches to the first part of the problem over the years have been the ways in which they have tackled the data gathering part of the exercise. Some have been so affected by the philosophy of the traditional examination question (that is that all the data needed is given and all the data given is needed) that they are quite outraged by the author's innocent assertion that he does not know, for instance, the specific heat of salt and perhaps they could find it out for themselves. Others, for similar reasons, resolutely struggle on assuming that since the specific heat of salt isn't given it must be possible to 'solve' the problem without it.

The second part of the problem raises interesting issues. It is usually possible, through discussion, to introduce into the students' thinking the idea of two limiting cases between which the real solution must fall. This, in itself, is an important concept in the solution of real world problems. It is often the case that, whilst the actual solution to a problem cannot be obtained or can only be obtained at undue and unrealistic cost, it can be bracketed sufficiently closely by approximations to the solution to answer the real question being asked. This is a good example of such a case.

Many groups have progressed to the point of attempting to formulate a set of equations similar to (11.4) or, equivalently, (11.5). Of those who have succeeded in deriving (more or less correctly) such a statement of the problem none, to the author's knowledge, have considered the possible analytic solution of the problem. This probably stems from a lack of breadth of knowledge of the standard solutions of the common partial differential equations of physics and engineering. Most groups have attempted some form of numerical solution. Unless the students have previously taken a course in the numerical solution of partial differential equations, or in finite difference methods, it is unusual for them to propose a solution by an implicit method. More usual is the use of a relatively simple explicit method. Such methods can be used on equations (11.5) or their equivalent but care must be taken to choose a combination of space and time discretisation interval which yields a convergent and stable form of the problem. A few students have finally succeeded in producing numerical solutions to what are effectively the two problems described by equations (11.5) and solved analytically by the Natural Gas engineers. It is satisfying to be able to report that students can produce solutions, using explicit finite difference techniques, which are, for all practical purposes, indistinguishable from the two curves of figure 11.4.

References

Abramowitz, M. and Stegun, I. A. (1965). *Handbook of Mathematical Functions*. Dover.

Agnew, J. L. and Keener, M. S. (1980). A case study course in applied mathematics using regional industries, *Amer. Math. Monthly*, **87**, 55.

Agnew, J. L. and Keener, M. S. (1981). Case studies in mathematics, *UMAP J*, **2**, 3.

Agnew, J. L., Keener, M. S. and Finney, R. L. (1983). Challenging applications: problems in the 'raw', *Maths Teacher*, **76**, 274.

Bajpai, A. C., Mustoe, L. R. and Walker, D. (1975). Mathematical education of engineers. Part I. A critical appraisal, *Int. J. Math. Educ. Sci. Technol.*, **6**, 361.

Berry, J. S., Burghes, D. N., Huntley, I. D., James, D. J. G. and Moscardini, A. O. (eds.) (1987). *Mathematical Modelling Courses*. Ellis Horwood.

Berry, J. S. and O'Shea, T. (1985). *Project Guide* (2nd edn.), from course materials for MST204 (Mathematical models and methods). Open University Press.

von Bertalanffy, L. (1973). *General Systems Theory*. Penguin.

Burghes, D. N. and Read, G. A. (1978). Mathematical modelling: editorial statement, *J. Math. Modelling Teachers*, **1**, 1.

Burkhardt, H. (1977). *Seven Sevens are Fifty? Mathematics for the Real World*. Inaugural lecture, Shell Centre for Mathematical Education, University of Nottingham.

Checkland, P. B. (1981). *Systems Thinking, Systems Practice*. John Wiley.

Churchman, C. W., Ackoff, R. L. and Arnoff, E. L. (1957). *Introduction to Operations Research*. John Wiley.

Clements, L. S. and Clements, R. R. (1978). The objectives and creation of a course of simulations/case studies for the teaching of Engineering Mathematics, *Int. J. Math. Educ. Sci. Technol.*, **9**, 97.

Clements, R. R. (1978). The role of simulations/case studies in teaching the practical application of mathematics, *Bull. IMA*, **14**, 295.

Clements, R. R. (1982a). Initial experience in the use of simulations/case studies in the teaching of Engineering Mathematics, *Int. J. Math. Educ. Sci. Technol,* **13**, 111.

Clements, R. R. (1982b). On the role of notation in the formulation of mathematical models, *Int. J. Math. Educ. Sci. Technol.,* **13**, 543.

Clements, R. R. (1982c). The development of methodologies of mathematical modelling, *Teaching Maths Applics,* **1**, 125.

Clements, R. R. (1984a). The use of the simulation and case study technique in the teaching of mathematical modelling. In *Teaching and Applying Mathematical Modelling,* ed. J. S. Berry, *et al.* Ellis Horwood.

Clements, R. R. (1984b). *BCSSP – User Manual.* Internal report, Department of Engineering Mathematics, University of Bristol.

Clements, R. R. (1985). *BSETR – User Manual.* Internal report, Department of Engineering Mathematics, University of Bristol.

Clements, R. R. (1986a). The role of system simulation programs in teaching applicable mathematics, *Int. J. Math. Educ. Sci. Technol.,* **17**, 553.

Clements, R. R. (1986b). Mathematical modelling using dynamic simulation, In *Mathematical Modelling Methodology, Models and Micros.* ed. J. S. Berry *et al.* Ellis Horwood.

Clements, R. R. (1989). Does overbooking pay? In *Case Studies in Mathematical Modelling,* vol. II, ed. I. D. Huntley and D. J. G. James. Oxford University Press.

Clements, R. R. and Wright, J. G. (1983). The use of guided reading in an Engineering Mathematics degree course, *Int. J. Math. Educ. Sci. Technol.,* **14**, 95.

Cornfield, G. C. (1977). The use of mathematics at the Electricity Council Research Centre in problems associated with the distribution of electricity, *Bull. IMA,* **13**, 13.

Cox, D. R. and Smith, W. L. (1961). *Queues.* Chapman and Hall.

Drew, D. A. (1981). Models of traffic flow. In *Case Studies in Mathematical Modelling,* ed. W. E. Boyce. Pitman.

Dym, C. L. and Ivy, E. S. (1980). *Principles of Mathematical Modelling.* Academic Press.

Ford, B. and Hall, G. G. (1970). Model building – a philosophy for applied mathematics, *Int. J. Math. Educ. Sci. Technol.,* **1**, 77.

Gaskell, R. E. and Klamkin, M. (1974). The industrial mathematician views his profession: a report of the committee on corporate members, *Amer. Math. Monthly,* **81**, 699.

Gradshteyn, I. S. and Ryzhik, I. M. (1965). *Tables of Integrals, Series and Products,* Translated by A. Jeffrey. Academic Press.

Greenberg, H. J. (1962). Applied mathematics: what is needed in research and education, *SIAM Review,* **4**, 297.

Griffiths, H. B. and McLone, R. R. (1971). Are academics fit to be trusted with the training of mathematicians?, *Bull. IMA,* **7**, 104.

Gross, D. and Harris, C. M. (1974). *Queuing Theory.* Wiley.

Hall, G. G. (1972). Modelling – a philosophy for applied mathematics, *Bull. IMA,* **8**, 226.

Hall, M. (1967). *Combinatorial Theory*. Blaisdell.

d'Inverno, R. A. and McLone, R. R. (1977). A modelling approach to traditional applied mathematics, *Math. Gaz.*, **61**, 92.

Kaye, G. W. C. and Laby, T. H. (1986). *Tables of Physical and Chemical Constants*, 15th edn. Longman.

Klamkin, M. (1971). On the ideal role of an industrial mathematician and its educational implications, *Amer. Math. Monthly*, **78**, 53.

Kuhn, T. S. (1962). *The Structure of Scientific Revolutions*. Chicago University Press.

Lighthill, J. (1979). The mathematical education of engineers, *Bull. IMA*, **15**, 89.

Lin, C. C. (1967). Objectives of applied mathematics education, *SIAM Review*, **9**, 293.

Lin, C. C. (1976). On the role of applied mathematics, *Advances in Math.*, **19**, 267.

Lin, C. C. (1978). Education of applied mathematicians, *SIAM Review*, **20**, 838.

McDonald, J. J. (1977). Introducing mathematical modelling to undergraduates, *Int. J. Math. Educ. Sci. Technol.*, **8**, 453.

McLone, R. R. (1971). On the relationship between curriculum, teaching and assessment of mathematics, *Int. J. Math. Educ. Sci. Technol.*, **2**, 341.

McLone, R. R. (1973). *The Training of Mathematicians*. SSRC Research Report.

McLone, R. R. (1976). Mathematical modelling – the art of applying mathematics. In *Mathematical Modelling*, ed. J. G. Andrews and R. R. McLone. Butterworth.

Moscardini, A. O., Cross, M. and Prior, D. E. (1984). On the use of simulation software in higher educational courses, In *Teaching and Applying Mathematical Modelling*, ed. J. S. Berry *et al*. Ellis Horwood.

O'Carroll, M. J. (1981). Performance of hydraulic buffers, In *Case Studies in Mathematical Modelling*, ed. R. Bradley, R. D. Gibson and M. Cross. Pentech Press.

Oke, K. H. (1980). Teaching and assessment of mathematical modelling in an MSc course in mathematical education, *Int. J. Math. Educ. Sci. Technol.*, **11**, 361.

Pollack, H. O. (1959). Mathematical research in the communications industry, *Pi Mu Epsilon J.*, 494.

Prager, W. (1972). Introductory remarks to a symposium on the future of applied mathematics, *Quart. Appl. Math.*, **30**, 1.

Penrose, O. (1978). How can we teach mathematical modelling?, *J. Math. Model. Teachers*, **1**, 31.

Stern, M. D. (1987). Consanguinity of witnesses: A mathematical analysis, *Teaching Maths. Applics.*, **6**, 79.

UMTC (1978). Possible models for guided reading courses, In *Teaching Methods for Undergraduate Mathematics*, Proceedings of the University Mathematics Teaching Conference 1977, Shell Centre for

Mathematical Education, University of Nottingham.

Usher, J. R. and Simmonds, D. G. (1987). Modern training of an industrial/commercial mathematician, *Teaching Maths Applics*, **6**, 27.

Weast, R. C. (1985). *CRC Handbook of Chemistry and Physics*, 66th edn. CRC Press.

Whitham, G. B. (1974). *Linear and Nonlinear Waves*. John Wiley.

Woods, L. C. (1969). What is wrong with applied mathematics, *Bull. IMA*, **5**, 70.

Subject index

adiabatic gas law, *see* gas law
air spring, 27
approximation, 74–5, 114–15, 130, 139

bicycle, 1, 4
buffer, hydraulic, 25–8

case studies, 2, 3, 46–8
case studies, assessment of, 56–8
case studies, creation of, 49, 50–2
case studies, mathematical content of, 51
case studies, objectives of, 46, 50
case studies, student reaction to, 54–6
case studies, use of, 3–4, 52–4
centrifugal pumping, 126
compound interest, 82
computer algebra, 40
constraints, mathematical, 33–7, 75
Crank–Nicholson, 156
curve fitting, least squares, 97–8, 101–2

data, accuracy of, 21, 27–8, 39, 130
data, incompleteness of, 39, 49, 101, 129,
  130
data, reciprocal principle of, 39, 44, 129,
  130, 157
data collection, 39, 129, 157
differential equation, 73, 127, 149, 150,
  152–4
diffusion equation, 149–56

elliptic integral, 36
energy, conservation of, 71
engineering mathematics, degree in, 42–3, 52
estimation, 27–8, 139–40
examinations, backwash effect of, 43, 44, 57

flowers, 38
four bar linkage, 72, 110

gas law, adiabatic, 25

groups, working in, 3, 46, 49–50, 53
guided reading, 1

human body, *see* body
hydraulics, 25–8

industry, use of mathematics in, 2, 6–10, 42,
  44–6
integration, numerical, 36, 73, 127, 131,
  154–6
iterative methodologies, 12–17
iterative refinement of models, 28–33

Laplace transform, 153–4
law, Jewish, 38

machine interference, 98–100
mathematical judgement, 14, 45, 55
mathematical modelling skills, need for,
  6–10, 43–6
mathematical modelling, learning, 4
mathematical modelling, methodology of,
  10–18, 20–3
mathematical modelling, nature of, 2, 7–9
mathematical modelling, non-mathematical
  content of, 44–5, 53
mathematical modelling, practical aspects of,
  24–40
mathematical modelling, teaching of, 48, 51,
  53–4, 56
mathematical modelling, tools for, 40
mathematics in industry, use of, *see* industry
mathematics teaching to engineers, 41
mathematics teaching, objectives of, 1, 46
mechanics, 1, 71–3, 110–13, 123–9
methodology of modelling, complex linked,
  20–3
methodology of modelling, iterative, 11–17
methodology of modelling, linear, 10–12
methodology of modelling, use of, 4, 23
methodology, systems, 18–20
models, working, 28

Newton–Raphson, 75
notation, importance of, 37–8
numerical integration, *see* integration

objectives of case studies, 46, 50
objectives of mathematics teaching, 1
operations research, 79–82, 97–101, 138–41

permutations, 38
problem, understanding the, 24–8

queueing theory, 98–100, 138–41

Runge-Kutta, 73

scientific method, 17–18

sensitivity analysis, 28, 39, 75, 129, 141–2
simulation, 40, 46–50, 127–8, 141
spreadsheet, 40
statistics, 40
systems methodologies, 18–20
systems theory, 18
systems, hard, 19
systems, soft, 19

tools, mathematical, 40
traffic lights, 29–33

viscous resistance, 127, 130–1

waiting time, 29–33, 99, 138–41
worst case analysis, 39, 114, 151–2, 157

Name index

Abramowitz, M., 36, 40, 154
Ackoff, R.L., 14
Agnew, J.L., 58
Arnoff, E.L., 14

Bajpai, A.C., 12–16
BCSSP, 40
Berry, J.S., 15, 58
von Bertalanffy, L., 18
Burghes, D.N., 16, 58
Burkhardt, H., 16

Carrier, G., 6, 7
Checkland, P.B., 18, 19, 23
Churchman, C.W., 14
Clements, L.S., 20, 58
Clements, R.R., 2, 16, 20, 23, 38, 40, 58
Cornfield, G.C., 45
Courant, R., 6
Cox, D.R., 98, 99
Cross, M., 40

Drew, D.A., 30
Dym, C.L., 17, 30

Electricity Research Council, 45
Engineering Mathematics (Department of), 41–3

Finney, R.L., 58
Ford, B., 10–12, 14

Gaskell, R.E., 9, 46
GINO, 40
Gradshteyn, I.S., 40
Greenberg, H.J., 6
Griffiths, H.B., 8
Gross, D., 138, 139

Hall, G.G., 10–15, 17
Hall, M., 38

Harris, C.M., 138, 139
Huntley, I.D., 58

Institute of Mathematics and Its Applications, 10
d'Inverno, R.A., 14, 15, 17
Ivy, E.S., 17, 30

James, D.J.G., 58

Kaye, G.W.C., 157
Keener, M.S., 58
Klamkin, M., 9, 10, 46
Kuhn, T.S., 18

Laby, T.H., 157
Lighthill, J., 10
Lin, C.C., 10–12

Mathematical Association of America, 9
McDonald, J.J., 14, 15
McLone, R.R., 8, 9, 14–17, 43, 46
Moscardini, A.O., 40, 58
Mustoe, L.R., 12–16

NAG, 40

O'Carroll, M.J., 25
O'Shea, T., 15
Oke, K.H., 58

Penrose, O., 15, 16
Pollack, H.O., 10
Prager, W., 7, 8
Prior, D.E., 40

Read, G.A., 16
Rosenbloom, 6, 7
Ryzhik, I.M., 40

SIAM, 6
Simmonds, D.G., 58

Smith, W.L., 98, 99
Stegun, I.A., 36, 40, 154
Stern, M.D., 38

UMAP, 10
UMTC, 2
University of Bristol, 2, 41, 53
Usher, J.R., 58

Walker, D., 12–16
Weast, R.C., 157
Whitham, G.B., 30
Woods, L.C., 10–12
Wright, J.G., 2

Yang, 6